AUSTRALIAN ENVIRONMENTS

The Murray-Darling Basin ..and The Coorong

ISBN: 978-1-7641194-4-3 (Paperback)

First edition: August 2025
A CONSERVATION CLUB PUBLICATION
www.conservationclub.com.au

A catalogue record for this book is available from the National Library of Australia

In the spirit of reconciliation, Conservation Club acknowledges the Traditional Custodians of Country throughout Australia and their connection to land, sea, and community. We pay our respects to their Elders, past and present, and extend that respect to all Aboriginal and Torres Strait Islander peoples.

To my grandchildren: Samuel, Jameson, Harry, and Audrey

..in the hope that they will enjoy sharing this journey of discovery ..

"And, above all, watch with glittering eyes the whole world around you because the greatest secrets are always hidden in the most unlikely places. Those who don't believe in magic will never find it."

Roald Dahl

CONTENTS:

REWARDS:

Conservation Club offers rewards (such as free downloads of colouring sheets and button badges) that can be earned by answering the 20 Question Quiz on page 175 and submitting one coloured-in image with the colours being truly representative of the species concerned. We encourage keen eyes, writing, recording, and reporting species sightings.

- **Buy one of our pocket notebooks! Designed to fit into the pocket of a child: at approximately A7 size, this blank notebook is an invaluable tool to have in order to record your observations of the environment you are exploring. It is really important to note the date, time of day, weather conditions and details of any sightings you may be lucky enough to experience! If the species is on the NCA, EPBC and/or IUCN Red List, please consider reporting your observations to an entity such as iNaturalist or Atlas of Living Australia! Citizen scientists are making great contributions to our understanding of these environments – you could be one of them!**
- **It is totally okay for you to guide your youngster and help them with this task (we suggest using the form at the back of the book as a guide before downloading the questions)!**

Email: conservationclubau@gmail.com for more information about how to apply for these rewards. The quiz questions are on the last page of this book. You can also join our mailing list to keep up to date with new releases.

Author's notes:

- Words and phrases shown within the book text in **BOLD** appear in the glossary/word lists at the back of this book.
- The whole point of the colouring pages is to encourage your young **conservationists** to engage with the book, tweak their observational skills, do a little bit of research, with the help of an adult if needed, and find out what colours are appropriate for the **species** concerned!
- Some of the words and language used may well be beyond what your young **conservationist** can understand, but it is suggested that you use the glossary to explain those words/phrases in simpler terms and at a level appropriate for your young wildlife warrior. They will ask for more information when they are ready!
- You can also choose to share the facts that you feel your young wildlife warriors are able to understand—you do not have to use them all! Just pick a few according to your child's age and comprehension skills.... You can always come back to the text later to expand their knowledge and vocabulary as they learn.

COLOURING RECOMMENDATIONS:

We recommend using colour pencils rather than texters or felt pens to ensure the information pages remain readable. Texters can bleed through the colouring side, affecting the readability of the fun facts!

We encourage adults and youngsters to find coloured images or photos online to guide the colouring. Familiarising both you and them will assist in being able to recognise the **species** if you sight them!

INTRODUCTION

In the Murray-Darling Basin .....it's mostly about water!

In the following pages, we will focus on **species** that occur in the **Murray-Darling Basin,** which occupies 14% of Australia's landmass. The **Murray-Darling Basin** is a large geographical area in the interior of south-eastern Australia, encompassing the drainage basin of the **tributaries** of the Murray River, Australia's longest river, and the Darling River, a right **tributary** of the Murray and Australia's third-longest river. The 'Basin', which includes six of Australia's seven longest rivers and covers around one-seventh of the Australian landmass, is one of the country's most significant **agricultural** areas, providing one-third of Australia's food supply. Located west of the Great Dividing Range, it drains southwest into the Great Australian Bight. It spans most of the states of New South Wales and Victoria, the Australian Capital Territory, and parts of the states of Queensland (the lower third) and South Australia (the southeastern corner).[2]

The Basin is 3,375 kilometres (2,097 miles) in length, with the Murray River being 2,508km (1,558 miles) long. Most of the 1,061,469km^2 km^2 Basin is flat, low-lying and far inland, and receives little direct rainfall. The many rivers it contains tend to be long and slow-flowing, and carry a volume of water that is large only by Australian standards. [2]

The way in which the water in the Murray River is shared is complex and is managed by the Murray-Darling Basin Authority, which calculates each state's share of the water before it can be allocated to farmers and other water users. Fortnightly updates are provided to advise the different states of how much water is available to them. These calculations are based on critical human needs, conveyance reserves, evaporation losses and flows to physically deliver water throughout the system. Not all the water in the Murray River system is shared between the three states. The following diagram is indicative of which states own the water flowing through them.

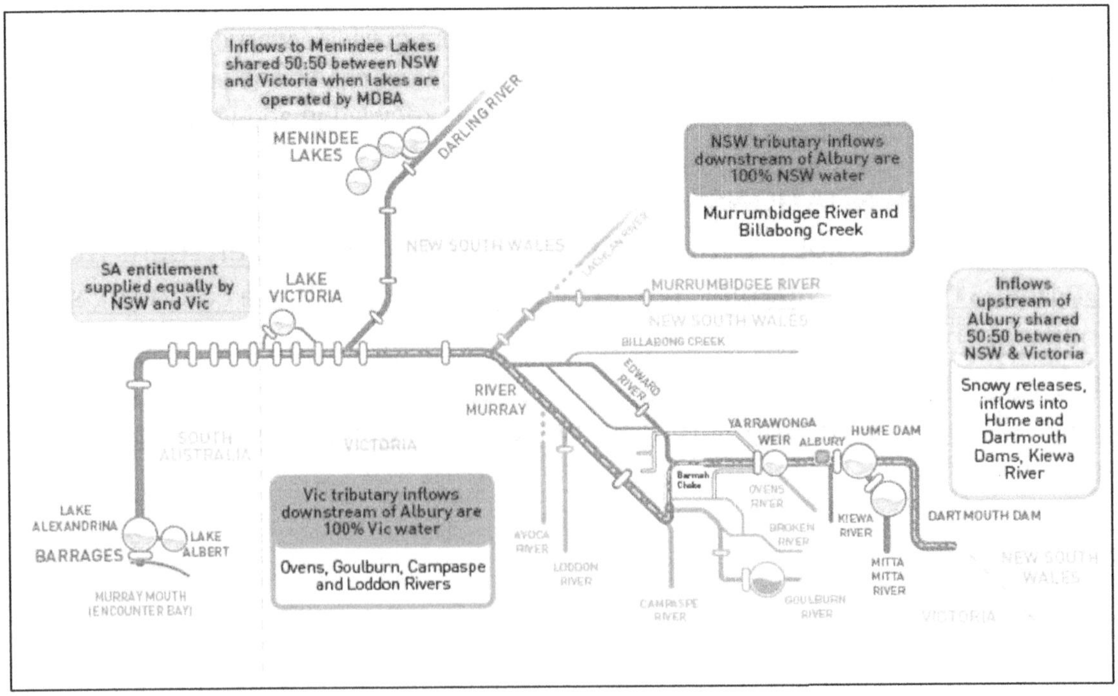

[3]

The Murray-Darling Basin

There are 23 rivers in the Basin, 77,000km of waterways stretching from mountain ranges into **arid** and **semi-arid** environments.

The Murray-Darling Basin holds significant environmental, **cultural**, and economic value for Australia. It is home to 16 internationally significant wetlands (although there are over 30,000 wetlands in total) and more than 2.4 million people (including people from more than 50 different First Nations). In addition to attracting visitors from around the world, which generates around $15 billion in tourism revenue each year, the Basin accounts for approximately 40% of Australia's **agricultural** produce. This includes 100% of our rice, 74% of our grapes, and 30% of our dairy.

From the 1950s, human populations and industries grew rapidly, and so did the amount of water taken from the river, resulting in less water for the environment. The rivers in the Basin became less healthy, especially during droughts. Reduced river flow meant there was more salt in the Basin's rivers, and more outbreaks of blue-green algae. When water levels dropped, acid sulphate soils were exposed, and blackwater events happened more often. There are fewer native fish, birds and mammals in the Basin than there were before Europeans arrived. At least 20 mammal species have become extinct, and conservation is needed for about half of the Basin's fish species.

There are about 15 bioregions partly or wholly in the Basin, including the mulga lands, Darling riverine plains, Murray-Darling depression, Riverina, NSW southwestern slopes and Australian Alps bioregions. These bioregions are divided into 419 subregions, which are differentiated on more specific aspects of geology, vegetation and other features that influence major regional ecosystems. Within each subregion, there are thousands of ecosystems. These ecosystems are highly **diverse**, ranging from wetlands and floodplains, forests and grasslands, alpine meadows and snowfields, to **semi-arid** deserts.

National parks and other reserves (which make up about 7% of the Basin's total area) protect many native plants and animals. Overall, though, only approximately 2.4% of the Murray-Darling Basin is protected, which includes areas protected under formal conservation reserves and national parks. The Basin also includes areas with informal protection through various management practices aimed at improving water quality and promoting sustainable resource use.

Information about some National Parks in the Murray-Darling Basin is given on pages 7-8.

The Coorong

All the rivers in the Murray-Darling Basin flow towards the Southern Ocean. Before reaching The Coorong, the Murray River passes through Lakes Alexandrina and Albert. These lakes were once part of a salt water **estuary** until five barrages were built to keep the fresh water in and the salt water out. These barrages ensure that the fresh water is available for farming, fishing and sailing; however, reducing the flow of the Murray River means that there is not enough flow to keep the mouth of the Murray open, and dredges have to be used continuously to allow the water to flow and to prevent the Coorong lagoon from becoming too salty.

The Coorong is a long, shallow, saltwater lagoon in South Australia, stretching over 100 kilometres (62 miles) and separated from the Southern Ocean by a narrow sand dune peninsula. It is part of the larger Coorong National Park, which covers an area of 490.15 square kilometres (189.2 square miles). The Coorong is recognised as a wetland of international importance under the Ramsar Convention.

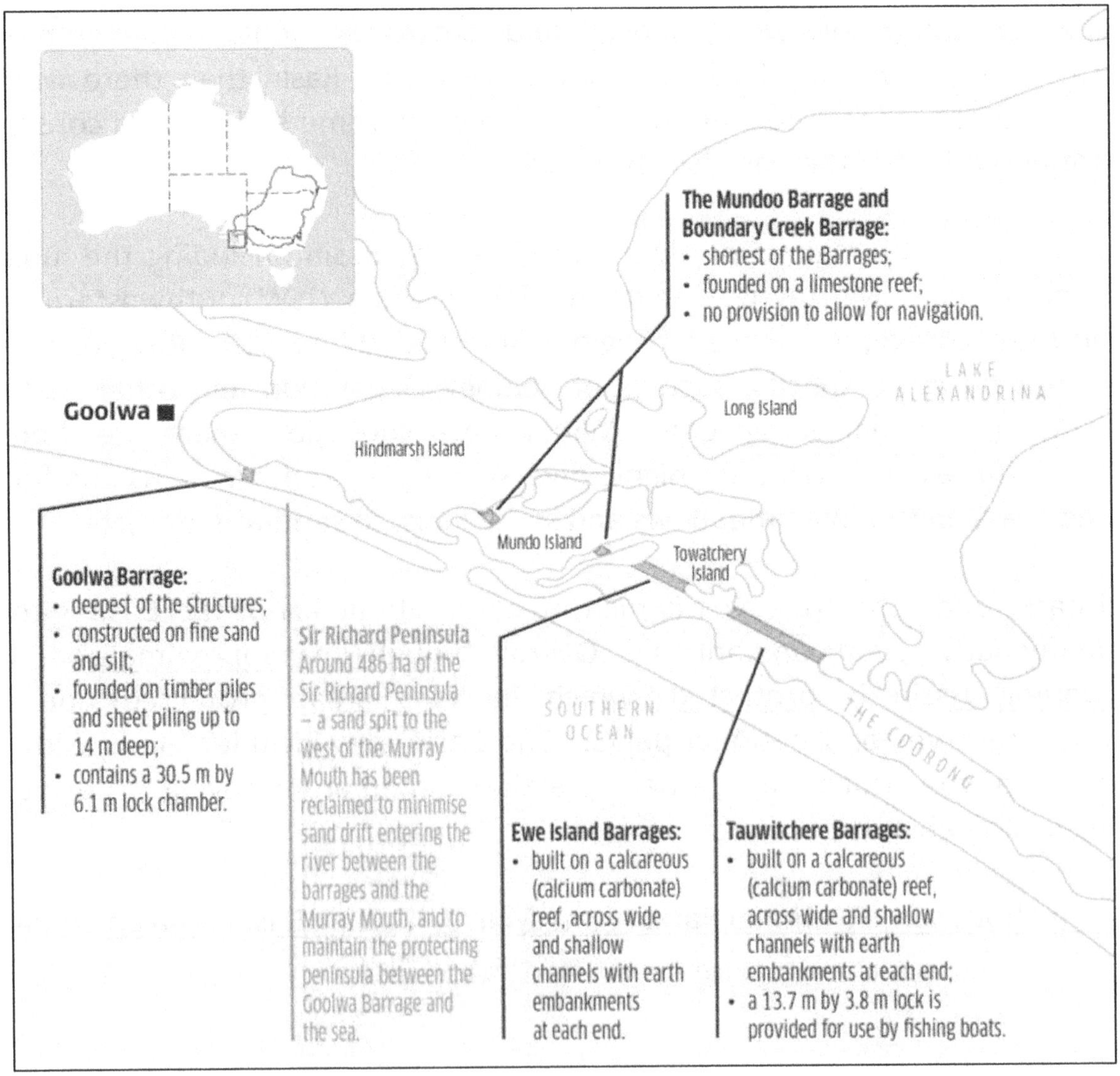

[4]

WILDLIFE IN THE MURRAY-DARLING BASIN

The wetlands and floodplains of the Murray–Darling Basin are home to a wide variety of native plants and animals, some of which are only found in particular stretches of the river. The Basin also supports a huge variety of birdlife. Some birds are native, while others are migratory and visit the Basin as part of their annual breeding cycle.

Among the native wildlife species of the Basin are 98 kinds of birds, 31 frog species, 46 varieties of snakes, 100 types of lizards, three species of freshwater turtles and over 50 species of fish.

Changes to the river system since **European settlement** have resulted in many **specie**s living in the Basin becoming endangered, despite conservation efforts.

The Murray-Darling Basin Authority states that the Basin is home to:

- 124 families of **macroinvertebrates**
- 46 snake species (5 endangered)
- Over 50 native fish species
- 31 native frog species
- 367 bird species (35 of which are endangered), including 98 species of waterbirds
- 85 mammal species (20 of which are extinct, 16 endangered)
- 3 freshwater turtle species
- 100 lizard species (1 endangered)

The Basin is estimated to be home to at least 779 different plant taxa. This includes a variety of trees, shrubs, and groundcover vegetation. Furthermore, 125 native wetland plant species have been identified across the wetlands in the southern part of the basin according to The Nature Conservancy Australia. The basin also supports 419 **riparian** plant species that grow along the edges of rivers, creeks, and wetlands.

The Murray-Darling Basin faces numerous threats to its **biodiversity**, primarily stemming from human activities and climate change. Key threats include habitat loss and **degradation**, unsustainable water management, the introduction of **invasive species**, and the impacts of climate change, such as increased droughts and altered river flows. These factors have a significant impact on the health of the basin's ecosystems and the survival of its native plants and animals.

DID YOU KNOW?

Australia is the sixth largest country on Earth and the only island/country that covers an entire **continent**! Because of its **isolation** from other **continents**, many of Australia's animal and plant **species** are found nowhere else. Australia also has a wide variety of **environments**, many of which are threatened for various reasons, including climate change, **extreme weather events**, and human impacts.

Around 80% of Australia's **native species** are **endemic,** and many are likely to become **extinct** within our lifetimes. Conservation Club AU focuses on raising awareness of **species** and **environments** in our 'backyards,' which we may or may not be familiar with. Based on **fossil** records, the natural **extinction** rate is approximately one **species** per one million **species** yearly. Scientists estimate that around 8.7 million **species** of plants and animals currently live on Earth. They are racing against time to catalogue Earth's **biodiversity** as **extinctions** continue. The Australian **continent** has seen more **mammal extinctions** than any other, along with **significant** losses of frogs, lizards, fish, insects, and plants. Many of these **extinctions** are recent. In Australia, there have been 21 likely **extinctions** of animals and four of plants, averaging about four per **decade**. Out of the 23 **extinctions** for which causes have been identified, 17 (73%) were primarily due to **invasive species**. Focusing just on animals, 17 out of 19 (89%) **extinctions** were mainly caused by **invasive species**. The four plant **extinctions** are primarily attributed to **habitat destruction** or **degradation**. The **primary culprits** in recent **extinctions** include the ***chytrid*** **fungus** (affecting **amphibians**)**, which has wiped out six **species**, and the wolf snake, which is responsible for the loss of four **species**. Other **significant** impacts include cats, foxes, black rats, brown trout, and a ***trypanosome*** blood **parasite**. Although **habitat** loss is often **cited** as the **primary** cause of **extinctions**, this does not hold true for recent **extinctions** in Australia. Earlier conclusions linked many plant **extinctions** to **habitat** loss. Still, increased efforts in searching have led to the rediscovery of many of these plants. The current data indicate that **invasive species** have been the leading cause of most **extinctions** in Australia since **European settlement.**

LET'S EXPLORE!

Kids in the 8 to 14-year-old age group are generally interested in spending time with their families exploring the great outdoors on weekend walks, camping getaways, etc. What a great way to spend time with your kids! It is so much better than burning up hours using electronic media and becoming **sedentary**, both of which can lead to behavioural problems and issues with sleep!

Kids also love to be part of a club and start collecting things they might be interested in, such as pebbles and shells. In my opinion, our kids are generally far more aware of **conservation** and **environmental** issues than in previous generations. This, therefore, presents an opportunity to educate, to get them interested, active, **motivated,** and involved in the **environment** and the **conservation** of **species** so that they will, hopefully, be able to work to preserve these animals so that they can show or tell their own children about them in years to come.

PLEASE…. GET OUT THERE and encourage these young and curious minds to engage with their **local environments** and the **critters** that can be found in them!

NATIONAL & MARINE PARKS:

Here are some national or marine parks you can visit within the Murray-Darling Basin. Because it is such a massive area of Australia and covers several states and a territory, this list is limited to just some of these protected areas. Still, for the keen explorer, there are many more to be found by searching on the internet in the states concerned.

QUEENSLAND:

For more information in Queensland, go to:
https://wetlandinfo.des.qld.gov.au/wetlands/facts-maps/aca-bounding-area-queensland-murray-darling-bulloo/

- Binya National Park
- Chesterton Range National Park
- Culgoa National Park
- Culgoa Floodplain National Park
- Currawinya National Park
- Girramay National Park
- Hell Hole Gorge National Park
- Lake Bindegolly National Park
- Mariala National Park
- Narkoola National Park
- Thrushton National Park

NEW SOUTH WALES:

- Cocoparra National Park
- Gundabooka National Park
- Kalyarr National Park
- Kinchega National Park
- Livingstone National Park
- Mungo National Park
- Murray Valley National Park
- Murrumbidgee Valley National Park
- Oolambeyan National Park
- Paroo-Darling National Park
- Willandra National Park
- Woomargama National Park
- Yanga National Park

VICTORIA:

- Barmah National Park
- Burrowa-Pine Mountain National Park
- Chiltern-Mt Pilot National Park
- Gunbower National Park
- Hattah-Kulkyne National Park
- Little Desert National Park
- Mount Granya State Park
- Mount Lawson State Park
- Murray-Sunset National Park
- Nyah-Vinifera Park
- Wyperfield National Park

SOUTH AUSTRALIA:

- Coorong National Park
- Murray River National Park – playing a vital role in conserving more than 13,000 hectares of wetlands and floodplains along the Murray River (divided into six sections: Katarapko, Lyrup Flats, Paringa Paddock, Kingston-on-Murray, Gurra Gurra and Bulyong Island)

ACT:

The ACT's national parks are not part of the Murray-Darling Basin's national park system.

Some background information about this book:

*The **species** in this book occur in the Murray-Darling Basin. Some of the **species** illustrated are either **Critically Endangered**, **Endangered**, or **Vulnerable** on the NCA and EPBC (SPRAT) lists.*

*On the IUCN Red List.[1], despite this being extensive, more than 169,420 **species** have been assessed at the time of writing this book. However, of the goal of 260,000 **species**, there are still 90,580 **species** remaining to be evaluated, an incredible achievement, but, for some **species**, due to the lack of up-to-date scientific data, this list is, at this point in time, not complete. The list is constantly updated, with a goal to reassess 142,000 **species**. More information can be found at https://www.iucnredlist.org/*

*The following tables give information about how the **species** are listed with NCA, EPBC (within Australia), and IUCN lists, within what categories, and a breakdown of the categories currently available for assessment.*

Younger readers may be unable to navigate this information themselves, so it is provided here to enable parents or caregivers to relay it to those young, curious minds!

I firmly believe that the age groups this book is aimed at (8 to 16 years) are curious and generally more informed than previous generations. They are aware that even one single voice can make a difference. This is evidenced by the number of young people taking up a cause. They recognise that, even one voice, one person, can make a difference, as evidenced by the number of young people taking up a cause.

The book is likely to appeal to a much broader audience than just the age group mentioned, as the illustrations and information are also appropriate for adults. Most importantly, we all need to be much more aware of our surroundings and report our observations to an entity such as SPRAT.

I sincerely hope that you, the Wildlife Warriors for The Murray-Darling Basin and The Coorong environments, will enjoy this book as much as I have enjoyed writing and illustrating it!

Elysia Rainwood

NCA – Status under the Nature Conservation Act 1992 (QLD legislation) [21]

EX	Extinct
PE	Extinct in the wild
CR	Critically endangered
E	Endangered
V	Vulnerable
NT	Near threatened
C	Least concern
SL	Special least concern
I	International
P	Prohibited

EPBC – Status under the Environment Protection and Biodiversity Conservation Act 1999 [2]

EX	Extinct
XW	Extinct in the wild
CE	Critically endangered
E	Endangered
V	Vulnerable
CD	Conservation dependent

IUCN – Status on the IUCN Red List [1]

EX	Extinct
EW	Extinct in the wild
CR	Critically endangered
EN	Endangered
VU	Vulnerable
NT	Near threatened
LC	Least concern
DD	Data deficient
NE	Not evaluated

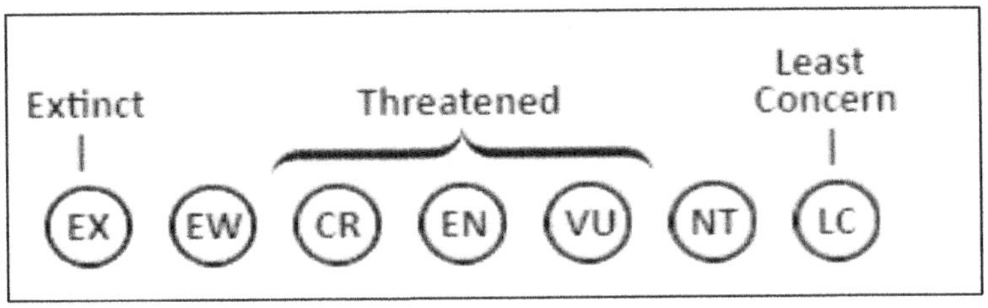

Where the species are on the **IUCN Red List,** the category is outlined in bold and grey in colour!

IUCN categories explained: [1]

CE = Critically Endangered:....... Facing an extremely high risk of extinction in the wild.

EN = Endangered:..................... There is a very high risk of extinction due to rapid population declines of 50 to more than 70% over the previous 10 years, a current population size of fewer than 250 individuals.

V = Vulnerable:......................... Being threatened with extinction unless the circumstances that are threatening its survival and reproduction improve.

- Sometimes the species on the IUCN Red List have not been listed simply because there has not been sufficient research to assess.
- In these cases, they are regarded as DD = Data Deficient **OR** NE = Not Evaluated.
- That does not mean that they are not critically endangered, endangered, vulnerable, threatened or any other classification... It just means that they have not been assessed!

Where on earth is the Murray-Darling Basin found?

Here is a map that helps to delineate the area of the Murray-Darling Basin:

[5]

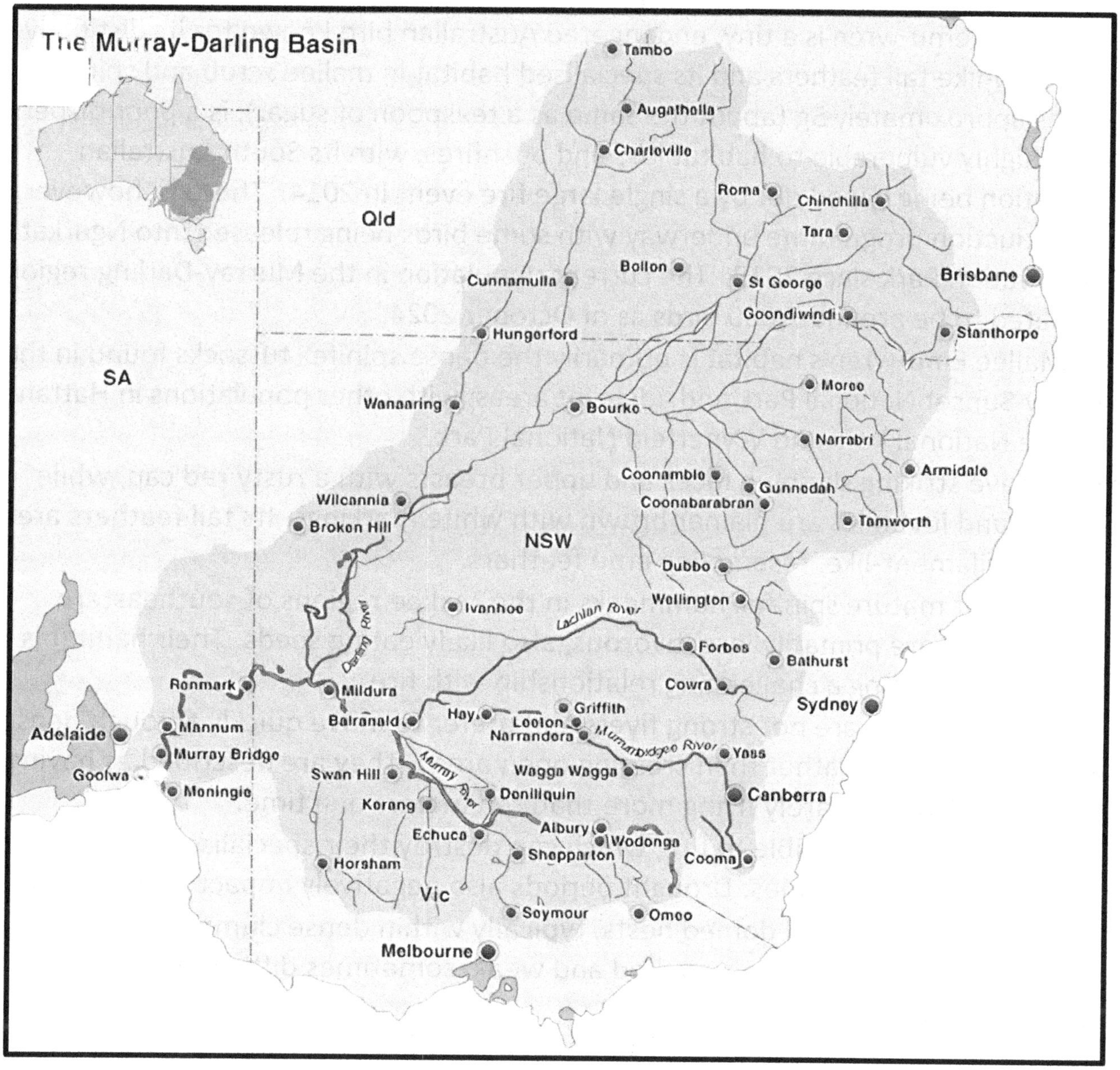

Mallee emu-wren

Stipiturus mallee

Vegetation illustrated: *Melaleuca thymifolia, Grevillea sp.*

- The Mallee emu-wren is a tiny, endangered Australian bird known for its distinctive, long, emu-like tail feathers and its specialised habitat in mallee scrub and spinifex. It weighs approximately 5g (about the same as a teaspoon of sugar), is a poor disperser, and is highly vulnerable to habitat loss and bushfires, with its South Australian population being wiped out by a single large fire event in 2014. There is, however, a reintroduction programme underway with some birds being released into Ngarkat Conservation Park since 2018. The current population in the Murray-Darling region is estimated to be around 7,000 birds as of October 2024.
- The Mallee Emu-wren's habitat is primarily the dense spinifex tussocks found in the Murray-Sunset National Park and adjacent areas, with other populations in Hattah-Kulkyne National Park and Wyperfeld National Park.
- Males have striking sky-blue faces and upper breasts with a rusty red cap, while females and juveniles are plainer brown with white markings. Its tail feathers are long and filament-like, resembling emu feathers.
- They inhabit mature spinifex hummocks in the Mallee regions of southeastern Australia and are primarily insectivorous, also likely eating seeds. Their habitat is very flammable, creating a challenging relationship with fire.
- Mallee emu-wrens are not strong flyers and prefer to move quickly through dense shrubs and spinifex rather than crossing open areas. They are described as having a "short flight" range, rarely flying more than ten meters at a time.
- They are highly susceptible to fire, which can destroy their specialised habitat and lead to population declines. Drought periods also negatively impact their numbers
- During spring, they build domed nests, typically within dense clumps of spinifex, and lay 2-3 eggs. Their call is high-pitched and weak, sometimes difficult to hear, especially in windy conditions.

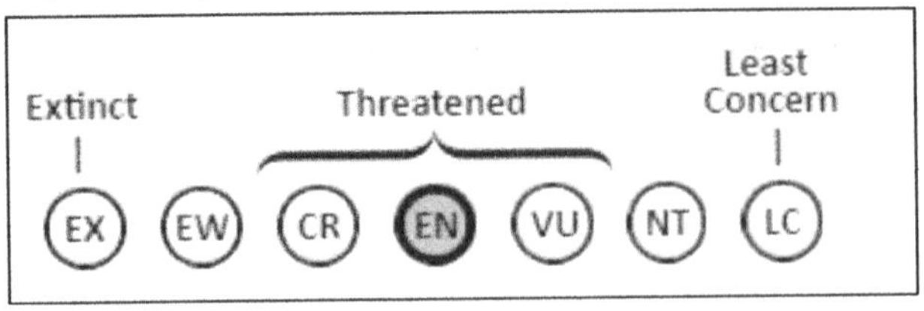

[1]

EPBC List = E = Endangered

Mistletoe bird

Dicaeum hirundinaceum

NATIVE TO MOST OF AUSTRALIA (ABSENT IN TASMANIA); ALSO SEEN IN THE EASTERN MALUKU ISLANDS OF INDONESIA, IN THE AROFURA SEA BETWEEN AUSTRALIA AND NEW GUINEA.

- The bottom-wiping mistletoe bird!
- The mistletoe bird is also known as the mistletoe flowerpecker, one of 54 species in the flowerpecker family, but is the only Australian example of the flowerpecker family.
- The mistletoe bird is highly adapted to its diet of mistletoe berries. It has a simple **digestive system**, through which the berries pass quickly, **digesting** the fleshy outer parts, meaning that the mistletoe seeds are still sticky. The bird is **distinctive** in that it sits on twigs and 'wipes its bottom' on the twigs, spreading the seeds and leaving them with enough **organic material** to grow a new plant quickly. In this way, the mistletoe bird ensures a constant supply of its main food source, although may also catch insects mainly to provide food for its young.
- As the name suggests, these birds are mistletoe feeding specialists, although the mistletoe itself does not need the birds to spread its seeds, as other fruit-eating birds, like the honeyeaters, also spread these.
- There are 92 species of mistletoe in Australia, more than anywhere else in the world, with about 70 being native and all 92 being endemic. Mistletoe is a **semi-parasitic** plant that needs its **host trees** for water but not necessarily for food. The plant can also **mimic** the appearance of the **host plants,** making it very difficult to tell which plant is the mistletoe and which is the **host plant**!
- Mistletoe birds are thought to have started settling in Australia about 2 million years ago.
- The mistletoe bird is small; 9–10 cm long and 7.5–11 g in weight. Males and females have different markings and colours in their **plumage**. The male mistletoe bird could be mistaken for a scarlet robin (*Petroica boodang*) or flame robin (*Petroica phoenicea*). The mistletoe bird is distinguished by its lack of a forehead spot and a black streak on its white belly. Its unique combination of size, shape, behaviour, and **plumage** makes it easily recognisable compared to other Australian **perching** birds, though it is closest in size and shape to the pardalote and thornbill. Their flight is fast, and they are usually seen flying above or high in the **tree canopy** on **distinctively** pointed wings.

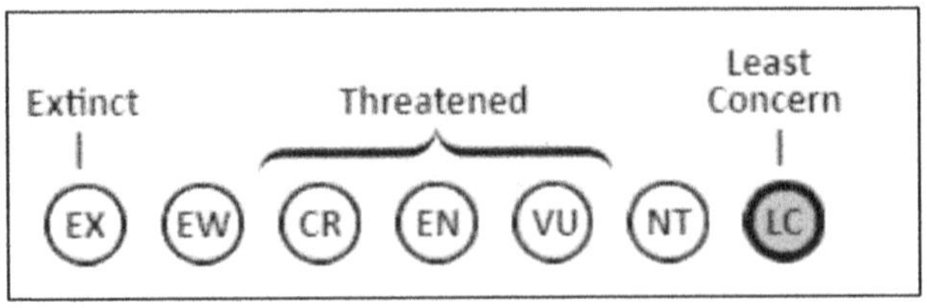

[1]

EPBC List = CE = Critically endangered

Metallic green carpenter bee

Xylocopa (Lestis) aeratus

- These bees are a large, iconic native bee species. The females are a beautiful jewel, metallic green in colour, and male *Lestis* carpenter bees have quite different colouring from the females, with a coat of fur over the **thorax**.
- They measure up to 17mm long, are eye-catching and make a loud buzz as they fly. They love to visit native pea flowers such as *Gompholobium*.
- They are friendly and harmless. The species is extinct on mainland South Australia and in Victoria, but still exists on Kangaroo Island, around Sydney and the Great Dividing Range in NSW. The last recorded sighting for Queensland was in 2019, south of Brisbane
- The bees went **extinct** on mainland South Australia in 1906, likely due to land clearing, and were last seen in Victoria in early December 1938 in the Grampians, which burnt completely during the Black Friday fires of January 1939.
- At the time of the 2020 fires, there were more than 150 nests in the stalks provided on Kangaroo Island, which contained mature brood. These all burnt. In NSW, the species has also been severely impacted, with much of its natural range burnt.
- Green carpenter bees make their nests in dead wood, which burns easily, making the bees vulnerable to fire. If their nests burn, adult female green carpenter bees might fly away but are unlikely to live long enough to reproduce again.
- Green carpenter bees mainly dig their nests in two types of soft wood: dry flowering stalks of grass trees and large, dead Banksia trunks. Grass trees flower prolifically after fire, but the dry stalks are only abundant between two and five years after fire. Banksia species don't survive fire and need to grow for at least 30 years to become large enough for green carpenter bees to use [6]. With increasingly frequent and intense fires, there isn't enough time for Banksia trunks to grow big enough before they're wiped out by the next fire.

The **Green Carpenter Bee Conservation Project** has, since 2013, been giving the bees alternative (artificial) nesting stalks to bridge the long periods of extreme low nest material. This nesting substrate is made of balsa wood and resembles dead flower stalks of grass trees or dead Banksia trunks. They are placed in the fire-affected areas where bees still occur. Female carpenter bees then excavate their own nest. Since the start, nearly 300 females have successfully raised offspring in these nests.

If you have seen a Green Carpenter Bee or a nest entrance in a dry grass tree stalk on Kangaroo Island, researchers would like to hear from you.

Please complete the details of your sighting on their form at
https://www.wheenbeefoundation.org.au/our-work/projects/green-carpenter-bee/sightings/

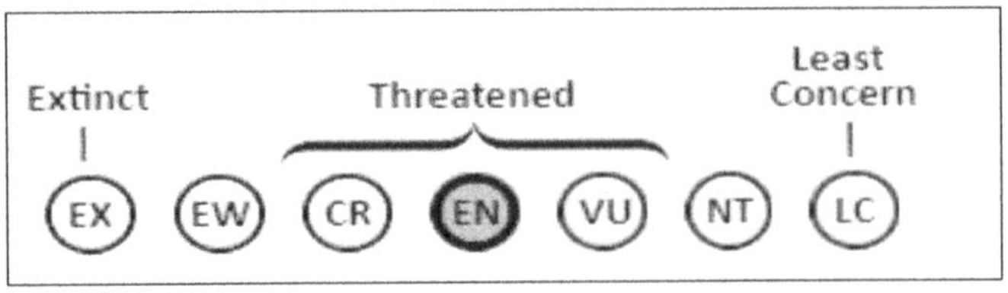

- ❖ **Environment Protection and Biodiversity Conservation Act – not currently listed – likely data deficient.**
- ❖ **NCA – not currently listed – likely data deficient.**

Masked owl

Tyto novaehollandiae novaehollandiae

- The Australian masked owl is a fascinating bird of **prey** with several notable **characteristics**. It's the second-largest **nocturnal raptor** in Australia, known for its distinctive facial disk and powerful talons.
- There are nine **recognised subspecies** of the Australian masked owl.
- Masked owls are among the largest owls in Australia, with a wingspan of up to 129cm and a body length of 47-51cm. Females are generally darker and larger than males.
- A **prominent**, heart-shaped facial disk with a dark ring is a key feature of this owl. They have large, dark eyes. They have well-feathered legs, large claws, and strong feet.
- They often have a stocky, crouched posture.
- Masked owls are **nocturnal**, hunting primarily in the early night hours. Females often hunt **prey** that is a lot larger than what the males hunt. This makes sense given that the females can be much bigger than the males.
- They are often found in timbered areas with a shrub understory. They nest in large tree hollows near **foraging** areas.
- Their diet includes small **mammals** like **rodents**, rabbits, bandicoots and possums. They also hunt possums, **reptiles**, birds, and insects.
- Like other raptors, these owls regurgitate pellets, which are soft clumps of **undigestible** bits of their **prey**, such as bones, claws, and fur. At one Australian masked-owl nest site, biologists found around 20kg of pellet material that had accumulated in and around the nest after many years of use![7]
- They are territorial and may remain in the same area once they have established a breeding **territory**.
- They usually lay two to three eggs. The female occupies the nest for up to 10 weeks before laying.
- They are **threatened** by loss of nesting **habitat**, clearing of **habitat** for development, and secondary **poisoning** from **rodenticides**. The masked owl population on the mainland is **declining**, with several states listing it as **threatened**.

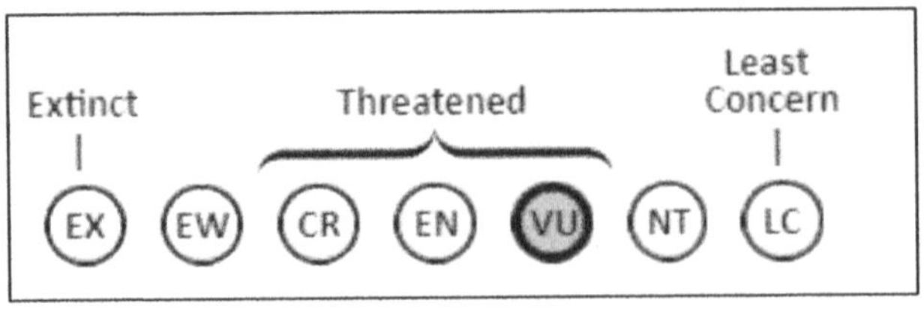

[1]

NCA List = LC* = Least Concern
*** At subspecies level this classification varies.**
EPBC List = V = Vulnerable

See page 66 (Powerful owl) for some interesting information about owls' eyes!

Agassiz's glassfish

Ambassis agassizii

NATIVE TO AUSTRALIA

- Agassiz's glassfish is **nocturnal**, hiding during the day and feeding at night – well, that's interesting! Who would have thought that some fish were **nocturnal**?
- Normally silvery-yellow to semi-transparent, a little bit olive/brown above with **dusky** scale margins.
- Threats include **predation** by non-native fish species such as Mosquito fish (*Gambusia holbrooki*), **degradation** of **habitat,** and all the other usual threats brought about by human impacts.

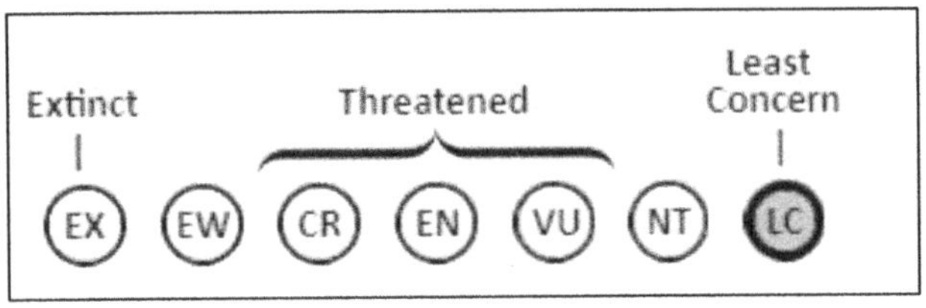

[1]

EPBC List = not categorised

Australian painted-snipe

Rostratula australis

- An endangered bird found in the wetlands and **endemic** to Australia.
- Sightings of Australian painted-snipes are so rare that even keen birdwatchers with many years working in the field forget they exist, and no one has recorded the call of these **elusive** waders. Scientists are trying to learn where they go when they disappear, which they do for months and even years at times.
- When spotted, they seem to have a knack of arriving when there is plenty of water or just in time to enjoy rainfall in some regions. So, how do they know where to go? So many questions can be asked about this mystery species. They feed on wetland **invertebrates** such as worms, **molluscs**, insects, and **crustaceans**, as well as seeds and other **vegetation**.
- They have excellent **camouflage** and easily hide in the mud or amongst water plants around wetland areas. They can drop into dips and keep one of their highly positioned eyes on their surroundings.
- It is believed there are only about 340 individuals left, but there are other things that also make them rare; little is known about them at all!
- They have been described as the wetland equivalent of the night parrot, which is so **elusive** that for 100 years it was thought to be **extinct**!
- A study assessing the **evolutionary uniqueness** and **conservation** status of nearly 10,000 bird species worldwide ranked the Australian painted-snipe at number 29, highlighting them as rare and unusual birds.

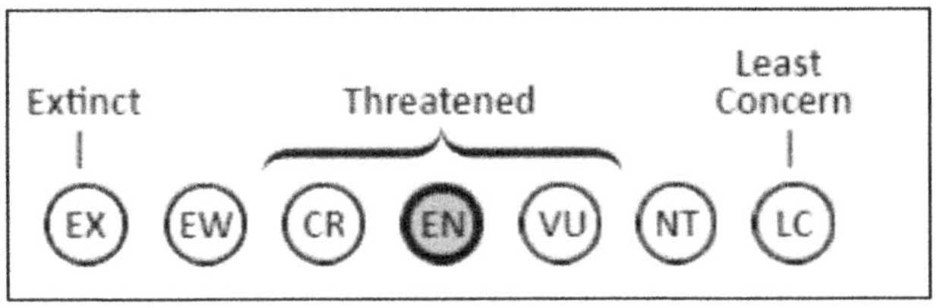

[1]

EPBC List = E = Endangered

10,000 Birds:

Have a look at this list, it is so interesting, and has regular updates, as well as emails, focusing on different species of birds all around the world! Why not subscribe to their email list to learn more about all sorts of birds in all sorts of environments?!

- ❖ Here is the website address: (https://www.10000birds.com/about)

Wandering albatross

Diomedea exulans

- Also known as the snowy albatross, the white-winged albatross and the goonie. These birds form part of the wandering albatross species complex, which has a complex and lengthy taxonomic history. You are unlikely to see these birds nesting around the mouth of the Murray River or in The Coorong; however, this area is still within their range.
- The wandering albatross is one of the two largest members of the genus *Diomedea*. It has the greatest known wingspan of any living bird and is also one of the most far-ranging birds.
- The genus, *Dimoedea,* is derived from Dimoedes from Greek mythology, whose companions turned to birds and *exulans* or *exsul* are Latin for "exile" or "wanderer" referring to its extensive flights.
- These large seabirds have a circumpolar range across the Southern Ocean. Some individuals are known to circumnavigate the Southern Ocean three times in one year, covering more than 120,000km.
- The plumage varies with age and also with the **subspecies**. **Juveniles** start as chocolate brown, but as they age, they become whiter. The adults have white bodies with black and white wings. The large **bill** is pink, as are the feet.
- They also have a salt gland that is situated above the nasal passage and helps desalinate their bodies, due to the high amount of ocean water that they imbibe. They **excrete** a high **saline** solution from their nose, which is a probable cause for the pink-yellow stain seen on the neck of some animals.
- Snowy albatrosses spend most of their life in flight, landing only to breed and feed. Distances travelled each year are difficult to measure, but one banded bird was recorded travelling 6,000 km in twelve days.
- When courting, they will spread their wings, wave their heads, and rap their bills together while braying. Wanderers have a large range of displays from screams and whistles to grunts and bill clapping. **(See illustration)
- Sailors used to capture the birds for their long wing bones, from which they made tobacco pipe stems. In the days of sail, the bird often accompanied ships for days, not merely following it, but wheeling in wide circles around it without ever being observed to land on the water.

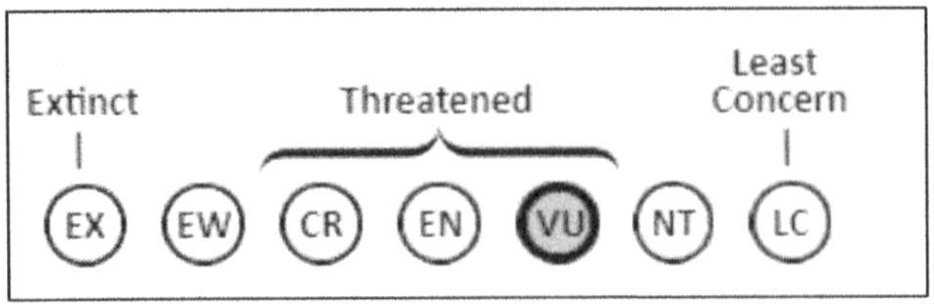

[1]

EPBC List = V = Vulnerable

Little Pygmy Possum

Cercartetus lepidus

- <u>The smallest possum in the world!</u>
- Although it looks like a dormouse **superficially**, it is the smallest of all known species of possum, ranging from 2.6-3mm in head-body length and weighing just 7-10g. Keep in mind that a teaspoon of sugar weighs about 5g so these little cuties weigh not much more than that! Measure out a teaspoon of sugar, and maybe add a further half teaspoon....WOW...so very tiny!
- Pygmy possums have long snouts with short whiskers, and their eyes are directed forward, surrounded by slightly darker fur. Their ears are mobile and largely hairless. The tail is **prehensile** and thickly furred at the base, which may be widened by fat stores beneath the skin. The remainder of the tail is relatively narrow and **cylindrical**, with only **sparse** hair between numerous tiny scales.
- The pygmy possum is **nocturnal** and **arboreal**. It lives primarily in shrubland or forest undergrowth, and, although a good climber, rarely ventures into the higher branches of trees, presumably because this would make it more vulnerable to avian **predators**.
- Pygmy possums use strips of bark to construct dome-like nests in tree cavities or rotten wood, but are **solitary** animals that do not share their nest with other individuals except for their own young.
- Pygmy possums are **omnivorous**, feeding on insects, spiders, small lizards, **nectar**, and **pollen**, the latter two primarily coming from Banksia and eucalypts. Their preference for eating **pollen** without destroying the host flower may mean that they help to **pollinate** some **species** of plant.
- Known **predators** include quolls, kookaburras, masked owls and tiger snakes.
- During colder weather, pygmy possums have the ability to enter **torpor.** While in this state, body temperature drops, and oxygen consumption falls to just 1% of normal.
- Breeding occurs throughout the year, although it may be more common in spring and summer. The female has a well-developed pouch containing four teats, which therefore limits the maximum size of a **litter** to this number. The young leave the pouch at around 42 days, although they may cling to the mother's fur and be carried about after this age. They leave the nest to fend for themselves at around 90 days of age.

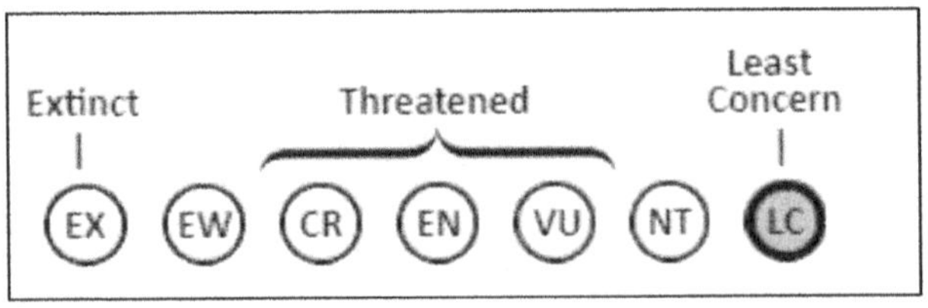

EPBC List = Not listed

River red gum

Eucalyptus camaldulensis

- Whilst the river red gum is not endangered in Australia, it is a **dominant** tree species of the Murray-Darling Basin and very important to the health of the Murray River as well as its tributaries, wetlands and floodplains.
- The red gum forests act as filters for the river, attracting and filtering water, and shifting and sieving nutrients throughout the floodplain landscape. The health of the Murray River and its river red gum forests are intrinsically linked – river channels and floodplains are part of the system.[8]
- The river red gum is so named for its brilliant red wood, which can range from a light pink through to almost black, depending on the age and weathering of the timber.
- These **iconic native** eucalypts usually grow to a height of 30m, although they can grow to 45m tall, and are thought to have a **lifespan** of up to 500-1000 years.
- They love water but can survive dry times. They can usually be found along rivers and creeks because they have special deep roots that can find water underground, even in times of drought. They can also survive being underwater for many months during floods.
- Older river red gums develop hollows in their trunks and branches, which are perfect homes for birds, small **mammals** like sugar gliders, and even insects.
- The limbs of river red gums, sometimes whole trees, often fall without warning so that camping or picnicking near them is dangerous, especially if a tree has dead limbs or the tree is under stress.
- Aboriginal peoples have used river red gums for shelter, food and medicine for a very long time, and they are **culturally significant** to many groups.
- The trees in the image are known as scarred trees or canoe trees[9]. These are trees that have had their bark removed by Aboriginal Australians for the creation of bark canoes. Bark was removed by making deep cuts in a tree with a stone pickaxe or other similar tool. The area of bark removed is typically regular in shape, often with parallel sides and slightly pointed or rounded ends, and the scar usually stops above ground level[9].

Australian fairy tern

Sternula nereis nereis

- There are three **subspecies** of fairy tern: the Australian fairy tern, the New Zealand fairy tern and the New Caledonian fairy tern.
- The Coorong, which is part of the Murray-Darling Basin, is a vital area for fairy terns, providing both foraging and nesting habitat.
- The Coorong is known to support breeding colonies of fairy terns, with successful breeding recorded in recent years.
- In 2024, there were 83 fairy tern chicks that successfully fledged from the breeding colony in The Coorong, according to BirdLife Australia. This was reported as one of the most successful breeding events at the site since monitoring began. The report also mentioned that 545 individual fairy terns were recorded in the Coorong, which was the highest count since 2001.
- The fairy tern is approximately 22–27cm in length, 70g in weight and has a wingspan of 44–53cm
- The fairy tern is much smaller than almost all other Australian terns.
- The fairy tern mainly feeds on fish, which it catches by hovering, beating its wings **rapidly**, 3 to 10m above the surface of the sea before diving, beak first, into the water. It also feeds on **crustaceans** and **molluscs**.
- The New Zealand **subspecies** has been on the brink of **extinction** for **decades**. With only around 40 birds left, the New Zealand fairy tern is one of New Zealand's most **endangered** birds.

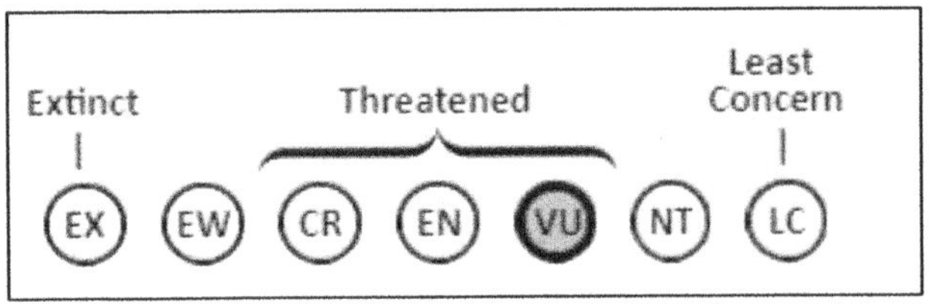

[1]

EPBC List = E = Endangered

Major Mitchell's cockatoo

Cacatua leadbeateri

- Major Mitchell's cockatoos are also known as pink cockatoos.
- They are one of the smaller cockatoos at around 36cm.
- They have a soft, salmon-pink body with a white crest that displays bands of scarlet and yellow when raised. The female has a broader yellow stripe on the crest and develops a red eye when mature. Males are usually bigger than the females.
- They inhabit the drier inland areas of Australia, with nearby water sources.
- It lives in scrublands, savannas, and wooded grasslands, but needs forested areas for its nesting habitat.
- They are omnivores, eating seeds, nuts, fruits, roots, and sometimes insects. They are known for their strong beaks used to crack seeds. They are particularly fond of Cypress pine and acacia seeds.
- They prefer to feed on the ground or among the branches of trees and shrubs.
- Major Mitchell's cockatoos are social birds. Outside of the breeding season, they spend time in groups and at night will occupy communal roosts. Flocks can number 10 to 50 individuals. They often can be seen in the company of other cockatoo species such as galahs and little corellas.
- They are weak fliers and have a characteristic slow, laboured flight at a low altitude. They will often fly for a short distance and rest before continuing their flight. They are a mainly sedentary species but will carry out local migrations seeking food.
- Unlike many other cockatoos, Major Mitchell's are very territorial during the breeding season, defending their nesting area from other pairs.
- Major Mitchell's cockatoos are monogamous, forming life-long pair bonds. In the mating season, the males attract female birds by strutting along branches while bobbing their heads up and down with raised crests. The breeding season is between August and October. Male and female both build their nest by gathering pebbles and bits of wood. The same nest can be used year after year. The female lays 2-5 eggs at the rate of one egg every 2 to 3 days. The incubation period is 23 to 30 days, with the young remaining in the nest for a period of 6 to 8 weeks before fledging[10]
- They can live for 50-60 years in the wild and even longer in captivity, sometimes exceeding 70 years.
- Their call is a loud, quavering "creek-ery-cree," which can be heard from a great distance.
- They are vulnerable to habitat loss due to land clearing and the illegal capture of birds for the pet trade.

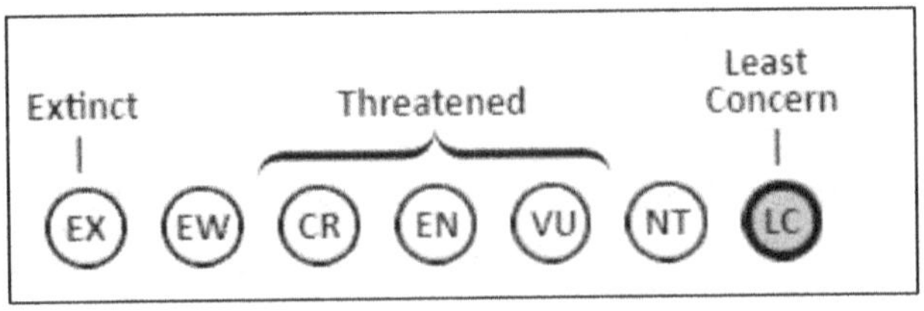

[1]

EPBC List = E = Endangered

Blue-winged parrot

Neophema chrysostoma

- The blue-winged parrot is a small, striking Australian parrot known for its distinctive blue and yellow **plumage**.
- Blue-winged parrots are found in various **habitats** across south-eastern Australia and Tasmania, including grasslands, grassy woodlands, and areas near wetlands. They can also be seen in altered **environments** like golf courses and airfields.
- Their population size is considered to be relatively low in certain regions. In Queensland, they are listed as vulnerable on both the EPBC and NCA lists.
- Many aspects of the movements of the blue-winged parrot are poorly understood. There is a definite passage of blue-winged parrots to and from Tasmania after breeding each year, leaving in March to April and returning in August to October. Some birds, however, overwinter in Tasmania or on the Bass Strait islands.[11]
- Primarily ground **foragers**, they feed on the seeds of **native** and **introduced** grasses, herbs, and shrubs, usually in pairs or small parties
- They form **monogamous** pairs and build nests in hollows of trees or stumps, preferably one with a vertical opening. The eggs are laid on a bed of decayed wood. The **clutch** size is four to six eggs, with the female **incubating** the eggs and both parents feeding the nestlings. Interestingly, it is thought that the clearing of forests may have benefited this **species** in terms of feeding, but not with regard to breeding, as potential nest sites are removed.
- Males have a distinctive two-toned blue band across the face, while females are duller with a less **prominent** band. Both sexes have olive-green upperparts, blue wings, and yellow bellies.
- They are closely related to other small Australian parrots like the rock parrot and elegant parrot, with some having similar colourations.
- Their call is described as a high-pitched, chiming sound, like a "zit zit". They may also make soft, twittering sounds while feeding. They sound more like a thornbill or fairy wren than a parrot.

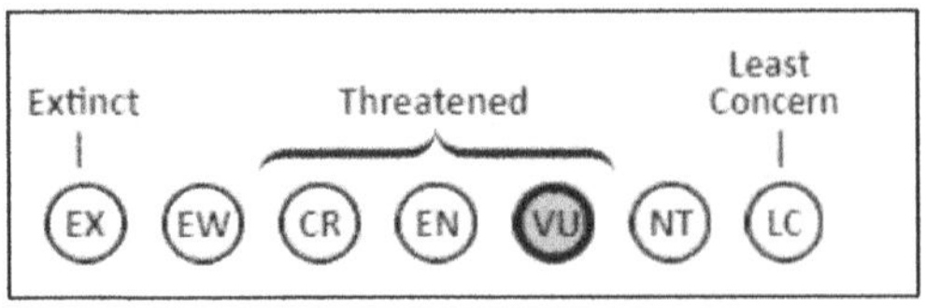

[1]

NCA List = V = Vulnerable
EPBC = V = Vulnerable

Common death adder

Acanthophis antarcticus

- Despite their name, common death adders are not related to adders, which belong to a different family, *Viperidae*.
- The common death adder's diet is mainly of frogs, birds, lizards, mice and rats. It uses its tail, the shape of a worm, which it wiggles to attract prey within striking distance. As **ambush predators**, they hide themselves, lying in wait, partially buried and well **camouflaged** in leaf litter, soil, or sand, with only their head and tail visible. Instead of actively hunting, they rely on this strategy to capture **prey**. The death adder is known for having one of the fastest strikes among animals, moving so quickly that the human eye cannot track it.
- They generally grow to be between 75 to 90cm in length.
- Common death adders normally take two to three years to reach adult size. Males are usually slightly smaller than females. Most have large bands around their bodies, though the colour itself is variable, depending on where they are found. Colours are usually black, grey or red and yellow, but also include brown and greenish-grey.
- Common death adders give birth to **litters** of live young, usually 3 to 20. However, over 30 young have been recorded in a single **litter**.
- Common death adders can inject, on average, 40–100mg of highly **toxic venom** with a bite. The **venom** of the common death adder attacks the nervous system, causing **paralysis,** which may seem minor but can cause death within six hours as the **respiratory system** shuts down. Before **antivenom** was introduced, it is reported that about 50% of death adder bites to humans were **fatal**.
- Threats include **habitat** loss and the spread of **invasive** cane toads, both of which are concerns. The cane toad eats young death adders, and the **toxic** glands on the toads' skin **poison** the adult death adders that eat the toads.

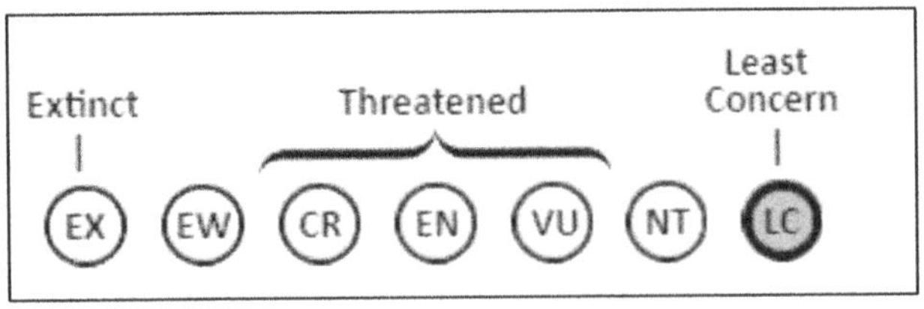

[1]

EPBC List = Not listed.

Diamond firetail

Stagonopleura guttata

NATIVE TO AUSTRALIA

- The diamond firetail has also been referred to as the spotted finch, spotted grossbeak, spotted-sided grossbeak.
- It is a small, stocky bird but one of the largest finches in Australia, weighing 15–19g and having a wingspan of 64–71 mm.
- The birds are very **distinctive,** with a black breastband on a white breast. The **flanks** are black with white spots, and the bird has a scarlet **rump** and a black tail. The **bill** and eye ring are coral red, and the legs and feet are dark grey.
- Diamond firetails generally eat ripe or partially ripe seeds found on the ground. They occasionally eat insects and their **larvae**.
- During courtship, the male diamond firetail carries a long piece of green grass in his beak. Then, he sits or **perches** on a branch near the female, bobbing up and down. When the female approaches, he twists his neck around and opens his **bill**, just like a young bird begging for food.
- The diamond firetail **constructs** its nest using green grass blades and stems, lining it with fine grasses and feathers. Both partners build the nest, but only the female does the weaving. These nests can be found in trees and shrubs with **dense** foliage. Birdlife Australia records "To safeguard their eggs and nestlings, diamond firetails are often recorded building their nests into the base of the large stick nest of a bird of **prey**, such as a whistling kite, white-bellied sea-eagle, wedge-tailed eagle, brown falcon, nankeen kestrel or a square-tailed kite. One nest of a whistling kite contained nine diamond firetail nests!"[12] Brave birds to be game to do that!
- Living in open grassy eucalypt forest and farmland, woodland, mallee country, **heath** and grassland with scattered trees, these birds stay mostly in one place.
- The diamond firetail's **habitat** is under **threat** due to changes in **vegetation** structure caused by factors like **overgrazing** and weed **invasion**, leading to a loss of **primary** food plants and **habitat**. This situation increases competition with **invasive species** and the risk of **predation**.

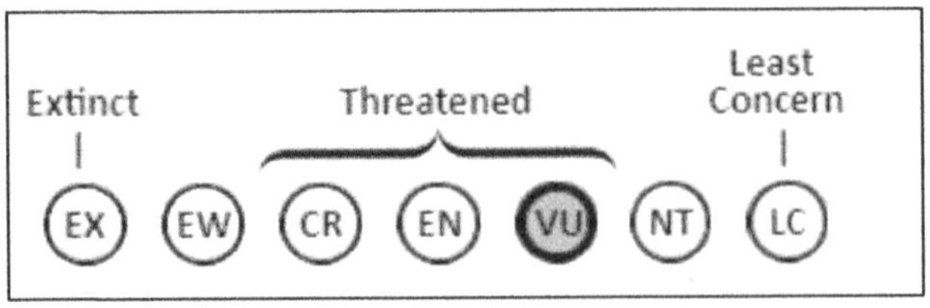

[1]

NCA List = V = Vulnerable

EPBC List = V = Vulnerable

Gudgeons

Carp gudgeons: *Hypseleotris sp.*

- There are a number of carp gudgeon species found in the Murray-Darling Basin, including the Southern Purple Spotted Gudgeon, the Carp Gudgeon (which includes multiple cryptic species like the Western carp gudgeon, Cryptic carp gudgeon, and Boofhead carp gudgeon), the Flathead Gudgeon, and the Dwarf Flat-headed Gudgeon.
- The gudgeon often swims among **aquatic** plants and branches, occasionally seeking shelter among rocks, which it uses as "caves". While some gudgeon species are **euryhaline**, meaning they can tolerate a wide range of salinities, others are strictly freshwater fish and cannot survive in saltwater.
- Gudgeons vary in size depending on the species, but generally range from 5 to 15cm in length. The firetail gudgeons are quite small, around 2 to 3cm. Some species lack a lateral line, while others do not. Typically, gudgeons have **ctenoid scales**.
- Their colouration varies, with some gudgeons developing brighter colours during the breeding season.
- Gudgeons are **omnivorous**, feeding mainly on small **crustaceans** and insects, but they also consume **algae** and other **aquatic** plants. Their diet includes a variety of **aquatic invertebrates** like water fleas and insect **larvae**, as well as **grazing** on **algae** and **detritus**.
- **Spawning** typically takes place during warm weather. After mating, the female lays rows of **adhesive**-coated eggs.These eggs **adhere** to various surfaces, such as plants, logs, rocks, or sand. The male then **fertilises** the eggs by releasing sperm into the water near the eggs.

Southern purple spotted gudgeon: *Mogurnda adspersa*

- The Southern Purple Spotted Gudgeon is an endangered** species within the Murray-Darling Basin. <u>CAN YOU FIND THE PURPLE SPOTTED GUDGEON IN THE ILLUSTRATION?</u>
- Southern purple spotted gudgeons are found in coastal streams of northern New South Wales and Queensland, as well as in the Murray-Darling Basin. Specifically, they inhabit slow-flowing or still waters such as rivers, creeks, and **billabongs**, often among **aquatic vegetation**, rocks, and snags.
- This species of gudgeon can reach up to 15cm in length.
- During breeding season, the southern purple spotted gudgeon exhibits vibrant colour changes.

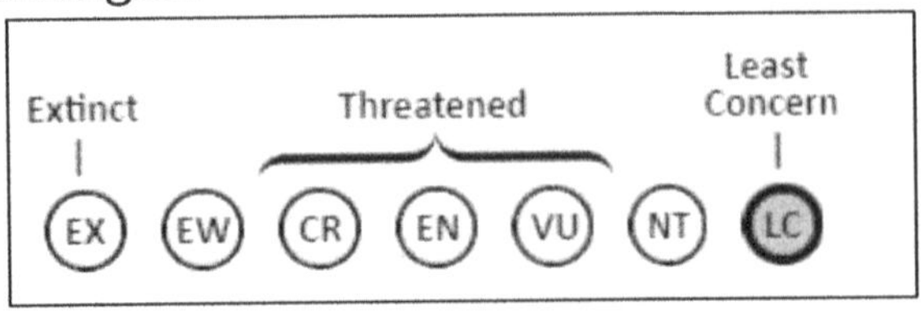

[1]

EPBC List = E = Endangered

** Southern purple spotted gudgeon

Yellow-footed rock wallaby

Petrogale xanthopus

- The yellow-footed rock wallaby is found in the **semi-arid** regions of eastern and southern Australia, including parts of the Murray-Darling Basin.
- This brightly coloured **marsupial** inhabits rocky outcrops, cliffs, and boulder fields, where it shelters in caves and **crevices** during the heat of the day and emerges to feed in the early morning and late afternoon.
- The yellow-footed rock wallaby is easily identified by its fawn-grey body, white underside, and distinctive rich orange to bright yellow forearms, hind limbs, and feet.
- In terms of **habitat**, the yellow-footed rock wallaby favours rocky areas like cliffs, gorges, and boulder piles within **semi-arid** woodlands. This broken country provides them with permanent waterholes and shelter sites from summer heat and **predator**s.
- Yellow-footed rock wallabies graze primarily on grasses, herbs, and low shrubs, obtaining water from their food.
- The yellow-footed rock wallaby is the largest rock-wallaby species in Australia, growing to a height of 80cm tall and weighing up to 11kg, with females being smaller at 60cm tall and up to 7kg.
- **Colonies** consist of breeding females with their offspring and a **dominant** male with a few **subdominant** males. Young females tend to stay at the **colony** site, while the **dominant** male forces young males out.
- Once found throughout the **semi-arid range** country in South Australia, New South Wales and south-west Queensland, the distribution and abundance of this species has declined dramatically, with the **species** now having a **fragmented distribution** of **isolated colonies**. This **specie**s is now generally restricted to protected areas that are actively managed to abate threats.
- Historically, intense hunting significantly reduced their numbers.
- Current threats include **predation** by introduced **species** like foxes and feral cats, and competition for food with domestic stock and **feral herbivores**, e.g. goats and rabbits.
- Conservation efforts include protection in reserves, captive breeding programs for reintroduction, and research into their **ecology** and threats.
- The species has played a crucial role in a cross-foster program for the critically endangered Brush-tailed Rock-wallaby.
- Females typically give birth to a single young, although twins are occasionally born. The **joey** develops in the mother's pouch for about 6 to 7 months before venturing out.

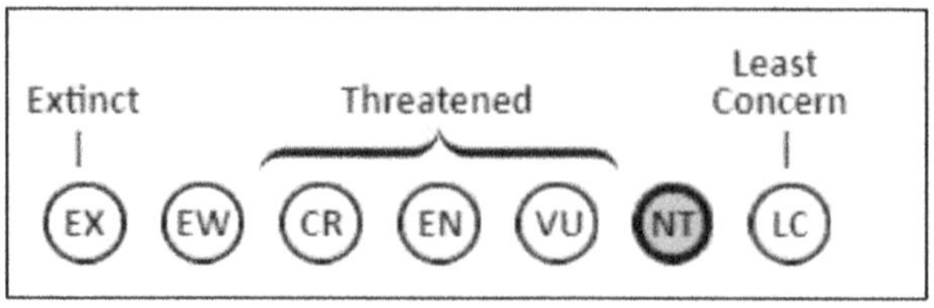

[1]

EPBC List = V = Vulnerable

Hairy pod wattle

Acacia glandulicarpa

- It is a dense, spreading shrub, typically 0.5 to 2 meters high, with dull to bright olive-green foliage and grey-brown stems.
- The leaves, called phyllodes, are thick, rigid, and oblong-elliptic, with a prominent central vein and a sticky, glandular texture, especially when young.
- The bright yellow, ball-shaped flower heads are simple, axillary, and contain 8 to 20 flowers.
- The distinctive seed pods are narrowly oblong, straight or curved, sticky, and covered in shining glandular hairs.
- It is endemic to south-eastern Australia, found in western Victoria and eastern South Australia.
- It grows in sandy and gravelly soils, often on slopes in the transitional zone between heavy and sandy soils.
- In its habitat, it is often found with species like Yellow Mallee (*Eucalyptus incrassata*) and Bull Mallee (*Eucalyptus behriana*).

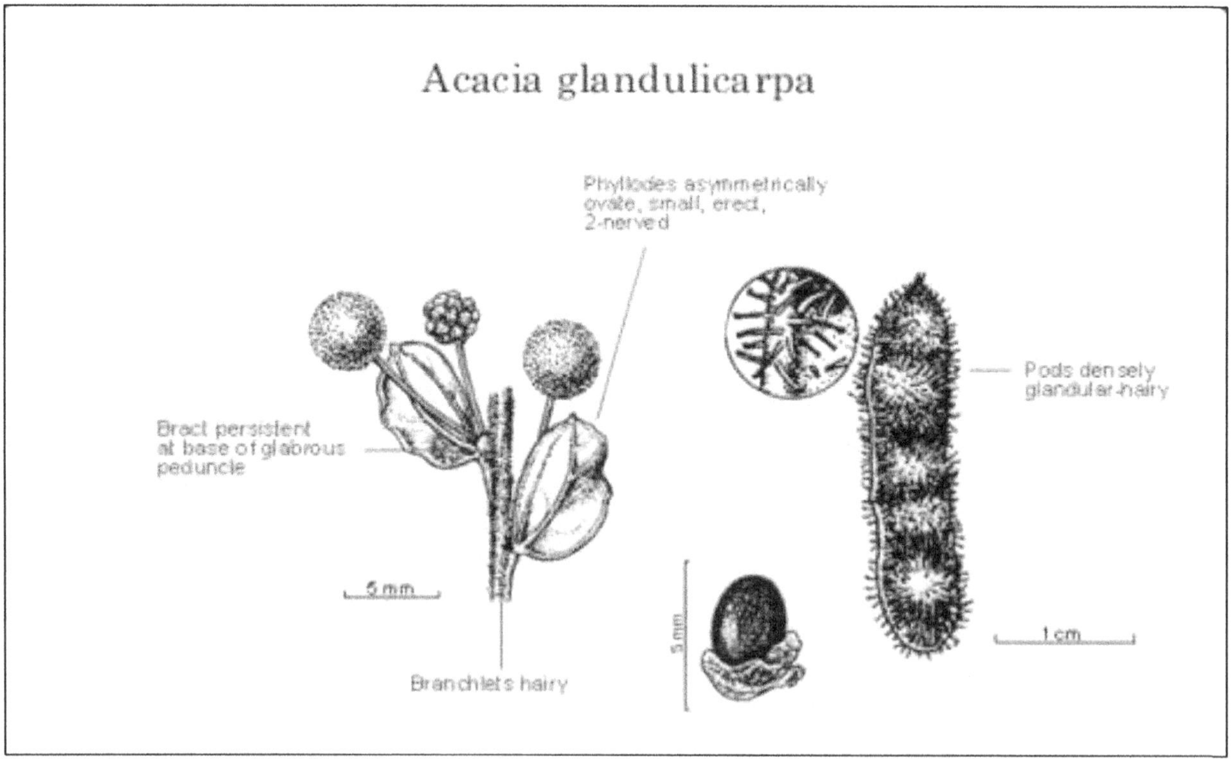

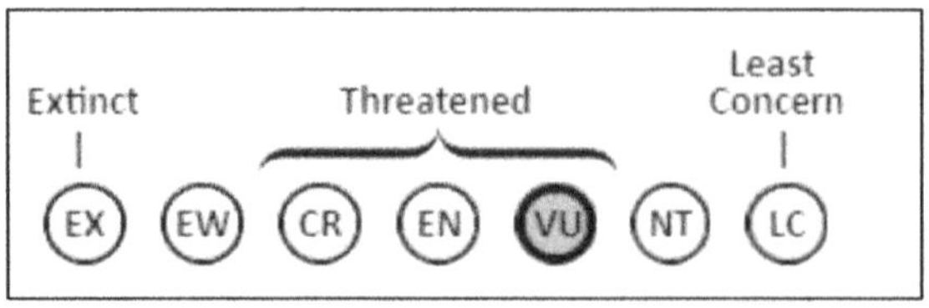

[1]

EPBC List = V = Vulnerable

Buffalo sallow-wattle

Acacia phlebophylla

- *Acacia phlebophylla* is a tree or shrub, growing up to 3m tall and occasionally reaching heights of up to 6m tall. Its **phyllodes** are **oblanceolate, obovate** or **elliptic** and commonly **asymmetrical** in shape, 4–14cm long and 1.5–6cm wide with **prominent** nerves and reticulated veins.
- Deep yellow rod-like flowers appear in June–December, widely scattered on 1-2 spikes per axil 4–7cm long. The fruits are straight or slightly curved **legumes**, 7–10cm long and appear in November–March, releasing 5–10 elliptical seeds, 5–7.5mm long.
- This species is known only from Mount Buffalo National Park, with an estimated 25 to 40 subpopulations. It mostly occurs in woodlands and heathlands on steep slopes between 500-1200m above sea level. It grows in shallow to skeletal soils of granite, often amongst boulders.
- *Acacia phlebophylla* is one of the purest natural sources of the **psychedelic drug dimethyltryptamine**, also known as DMT, which occurs as the predominant **alkaloid** throughout the plant

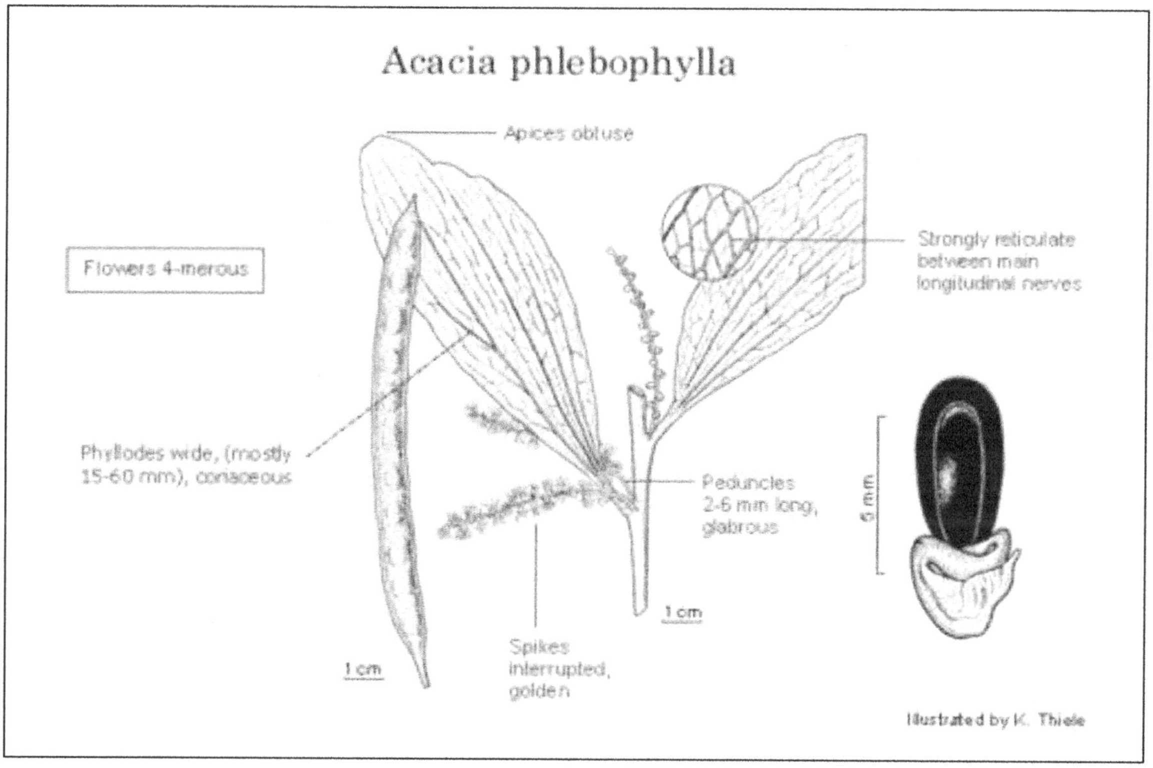

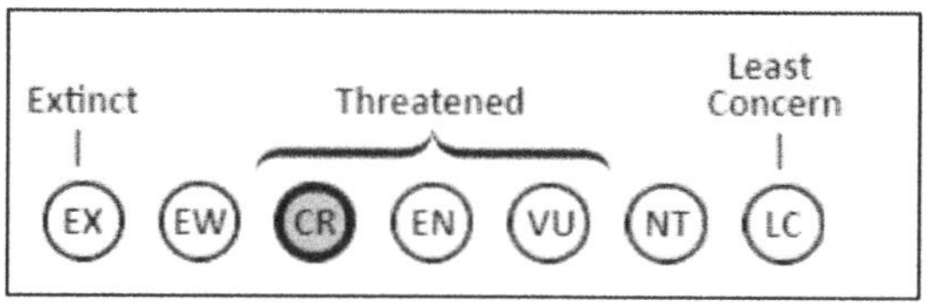

[1]

EPBC List = Not currently listed

Conservation status:
VIC = CE = Critically endangered

Superb parrot

Polytelis swainsonii

- The superb parrot is a beautiful and unique Australian bird, known for its vibrant colours, particularly in males, and its preference for woodland habitats near water.
- Male superb parrots are bright green with a yellow forehead and cheeks, and a striking red band across their chest. Females are primarily green, but their faces are a lighter blue-green; their flight feathers are a bit darker. They are about 40cm long and have long, pointed, elegant tails.
- Adult male superb parrots have bright yellow on their foreheads and throats. They also have a red stripe across their throats.
- They love to eat flowers, fruits, **nectar**, and **pollen** from eucalyptus trees, and they also enjoy seeds from grasses and plants. They also eat seeds and green parts of the Yellow Box tree (*Eucalyptus melliodora).*
- They often find their food on the ground in the early morning and late afternoon.
- Superb parrots build their nests in hollows or hollow limbs of large trees, often near water.
- They can live in small **flocks** or families. Sometimes, several pairs of parrots will nest close to each other, forming a small group of up to six pairs. This means the females are also close by.
- These parrots can live for up to 25 years or more.
- The superb parrot was once listed as a vulnerable species on the IUCN Red List of Threatened Species; this meant it was at risk of becoming endangered. However, in 2012, its status was changed to least concern. This is good news, but it still means we need to protect them. The parrot is also listed on Appendix II of CITES, which helps control trade of the species.
- Superb parrots are found only in Australia, mainly in the dry woodlands of New South Wales and Victoria. In 2005, logging in the Barmah State Forest destroyed many nesting areas. This was a big problem, especially since fewer than 150 superb parrots were breeding in Victoria at that time.

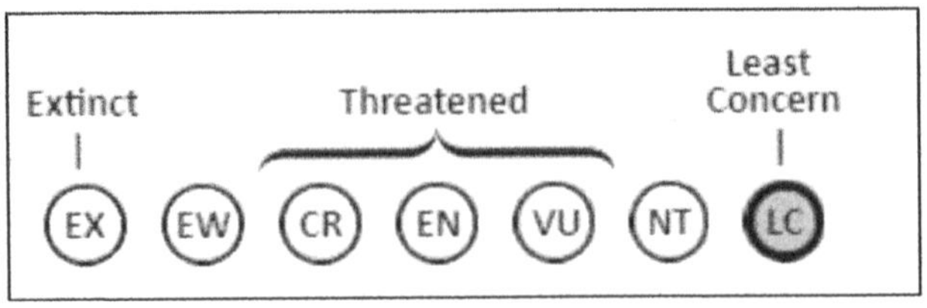

[1]

EPBC List = V = Vulnerable (NSW, ACT)
EPBC List = T = Threatened (VIC)

Curlew sandpiper

Calidris ferruginea

- The curlew sandpiper is a medium-sized wading shorebird with a long black bill, curved downwards at the end, and black feet and legs.
- This **distinctive** beak is what **distinguishes** the curlew sandpiper from other sandpipers of a similar size.
- Although their English name includes the word 'curlew', curlew sandpipers (*Calidris ferruginea*) are not actually curlews!
- Other common names are pygmy curlew, curlew stint and redcrop.
- They are 18-23cm long with a 38-41cm wingspan.
- The curlew sandpiper breeds on the **tundra** of Arctic Siberia. It is a **migratory** shorebird that spends winters mainly in Africa but also in Australia, New Zealand, and southeast Asia, flying thousands of miles. During **migration**, it has been recorded to fly for more than 48 hours without stopping.
- During breeding periods, its **plumage** is a rusty red/brown colour.
- Both males and females **incubate** the nests, which usually have 3-4 eggs.
- They are small wading birds, often seen in flocks with other waders.
- They **forage** in soft mud in sheltered coastal areas, mangroves, **estuaries**, lagoons, and marshes, eating mostly snails, prawns/shrimp, insects, and **polychaete worms**. They can be seen bobbing up and down at the water's edge, picking up and **probing** shallowly for prey.
- Although widespread, **populations** appear to have decreased dramatically in the last **decade,** and the IUCN Red List has judged this species as 'near threatened'.
- **Predators** include the Arctic fox, which will hunt waders when lemmings are not available or in poor supply.
- Other than **predators**, threats include human impacts such as development, **recreational** activities (4WD, walking, fishing, surfing, boating, etc.) at **foraging** sites, and **groundwater** pollution.

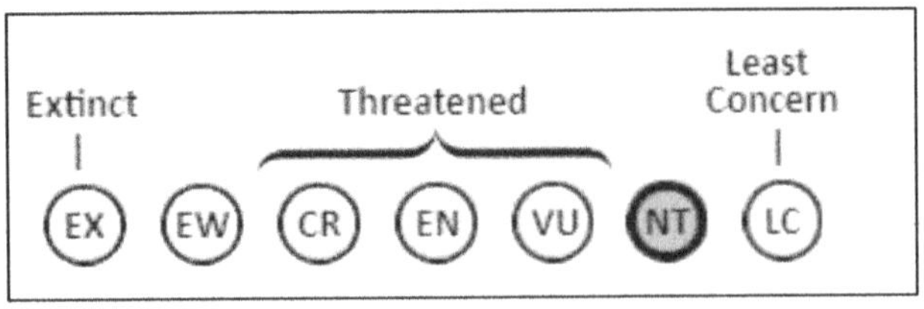

[1]

NCA List = CR = Critically Endangered (Qld)

EPBC List = CE = Critically endangered

Glossy black cockatoo

Calyptorhynchus lathami lathami

NATIVE TO AUSTRALIA

- Male and female glossy black cockatoos have different appearances. The word for this is **dimorphic**!
- They are the smaller of the two species in this **genus**.
- "Glossies" pair for life.
- They are one of Australia's rarest birds.
- They are **social** creatures although usually seen in pairs or small groups but rarely form large flocks.
- **Sheoak** seeds, which are inside the cones of the **sheoak** trees, are their **primary** food source. Still, they sometimes eat other things, such as the 'Chonky apple', also known as the Chine apple (*Zizphus mauritiana*). I observed this first-hand when living in Karumba many years ago... When the 'Chonky apples' were fruiting, driving past these small, spreading trees would disturb the glossies, and they would fly up in a blur of red/orange/black feathers. Quite a sight!
- They can be noisy, but they are usually very quiet when eating! <u>A good rule for good manners is to not talk with your mouth full!</u>

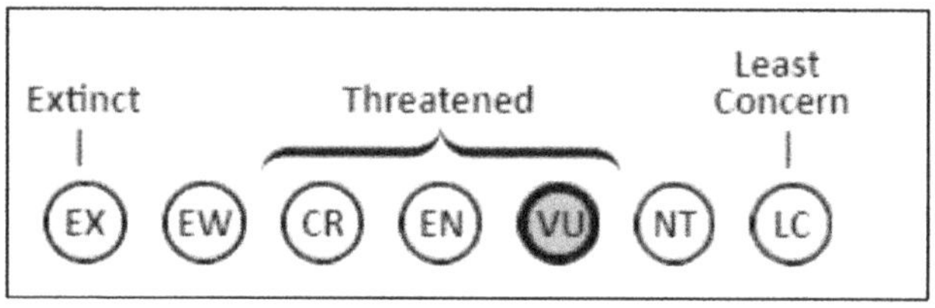

[1]

NCA = V = Vulnerable

EPBC = V = Vulnerable

Squatter pigeon (southern subspecies) **LOCALLY EXTINCT**

Geophaps scripta scripta

- The squatter pigeon historically occurred in parts of the Murray-Darling Basin but has since disappeared from its Southern range.
- Squatter pigeons are **unique** ground-dwelling birds with interesting features and behaviours. They are known for their distinctive black and white facial stripes, gray-brown **plumage**, and dark brown wings with some **iridescent** patches. The southern squatter pigeon is slightly larger than the northern (*Geophaps scripta*) **subspecies**, with blue-grey skin around its eyes, while the northern subspecies has orange-red skin.
- They are a medium-sized pigeon, typically around 30cm long and weighing between 190-250 grams.
- They also have a limited vocal repertoire, primarily using faint, mellow coos and plaintive crooning for communication.
- The southern squatter pigeon lives in arid and semi-arid areas, including grasslands and dry **sclerophyll forests**.
- Unlike many pigeons, squatter pigeons are primarily ground-dwelling birds, with a preference for **foraging** for food on the ground among sparse grasses.
- They feed on seeds, with some insects, particularly during the wet season. They **consume** seeds from both **native** and **introduced** plants, with grasses and legumes being particularly important.
- Squatter pigeons are reluctant fliers and often run away from danger rather than flying, displaying many sudden direction changes.
- They are usually seen in pairs or small **flocks**, but aggregations of over 30 have been reported.
- Squatter pigeons **roost** in low trees at night.
- Breeding in the squatter pigeon is poorly understood but appears to occur throughout the year and is likely influenced by **local** conditions. They breed as **solitary** pairs and may produce two **broods** in a season.

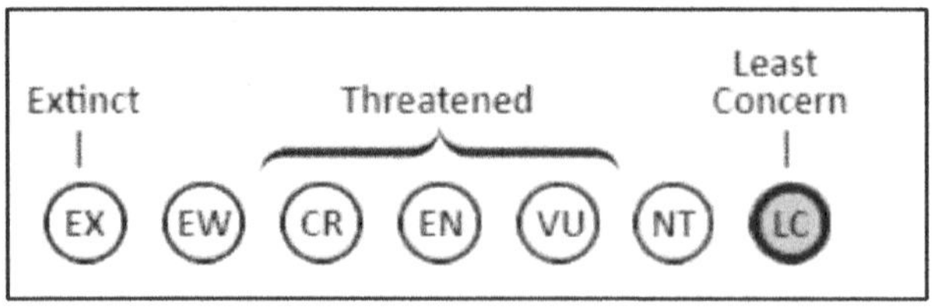

NCA List = V = Vulnerable
NCA NSW List= CR = Critically endangered
EPBC List = V = Vulnerable

[1]

Northern (*Geophaps scripta*) **subspecies**

Short-beaked echidna

Tachyglossus aculeatus

- Echidnas are **monotremes**, egg laying **mammals**, in the same order as platypus.
- The echidna has spines similar to a porcupine, a beak resembling a bird's, a pouch like a kangaroo, and it lays eggs like a **reptile**! Also called spiny anteaters, these small, **solitary mammals** are native to Australia, Tasmania, and New Guinea. Typically, they measure between 30-43cm in length and weigh between 1.8-4kg.
- The echidna usually lives in open heathlands, forests, woodlands, scrublands, and grasslands, often among **vegetation** or inside hollow logs. During poor weather, they seek shelter under bushes or burrow into the soil.
- Echidnas have a long nose (sometimes called a beak), small eyes and tiny faces. They don't have very good vision but have really good hearing and smell. They are also strong diggers, thanks to their long, rear-facing claws.
- Surprisingly, the spines on an echidna are an enlarged, tough form of hair; how about that?! As well as having spines, echidnas are covered in shorter fur that helps keep them warm. In colder climates echidnas have fewer spines and thicker fur.
- Their spines also serve as the echidna's main defence against **predators**. If threatened, they roll into a ball with spines pointing outward or dig themselves to safety.
- Short-beaked echidnas eat ants and termites.
- Echidnas do not have any teeth. They have slender snouts and tiny mouths. They use their long and sticky tongues to feed on worms, termites, ants, and insect **larvae**. Their 15cm tongues rapidly dart in and out of their mouths to slurp up insects. They break down their food using hard pads located on the roof of their mouths and the backs of their tongues
- Echidna milk is pink because of the high iron content.
- Female echidnas do not have nipples but have special glands in their pouches that secrete the milk they feed to their young, and the baby echidnas called '**puggles**', lap it up.
- During the breeding season, echidnas form lines known as 'mating trains.' The male echidnas line up, nose to tail, behind a single female, forming a train of up to a dozen individuals.

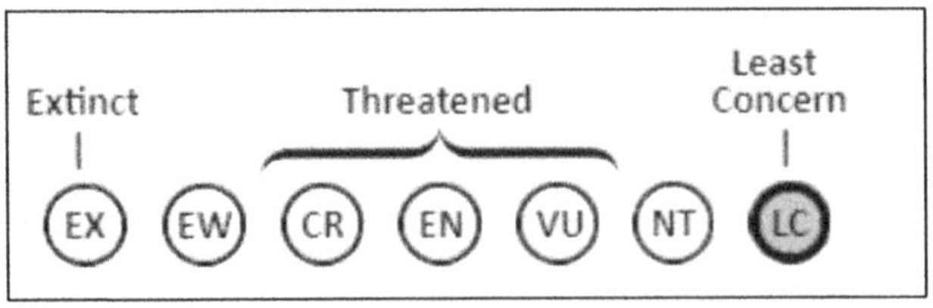

[1]

EPBC List = Not listed for mainland Australia but = E = Endangered for Kangaroo Island echidna

Regent honeyeater

Anthochaera phrygia

NATIVE TO AUSTRALIA

- The scientific name comes from the ancient Greek *Anthos*, "flower, bloom," and *khairo*, "enjoy," and the **specific species** name from Phrygius, a reference to the people of Phrygia (now Asian Turkey) who were skilled in the use of gold in embroidery.
- The regent honeyeater's diet consists mainly of **nectar** from Eucalyptus and mistletoe **species**. Sometimes they feed on insects and their **honeydew** as well as **native** and **cultivated** fruit.
- Only about 250-300 birds are **estimated** to remain in the wild.
- They are great **mimics** of other birds' songs but have a **unique** and **complex** song of their own. Research suggests that part of the reason for the **rapid decline** in **population** is that the young are struggling to learn the mating calls from the adult birds, which may also reduce their ability to find a mate and to breed successfully.
- **Captive breeding** programs are in place in New South Wales.
- The Pilliga woodlands are known to be an important area for breeding and foraging for regent honeyeaters.

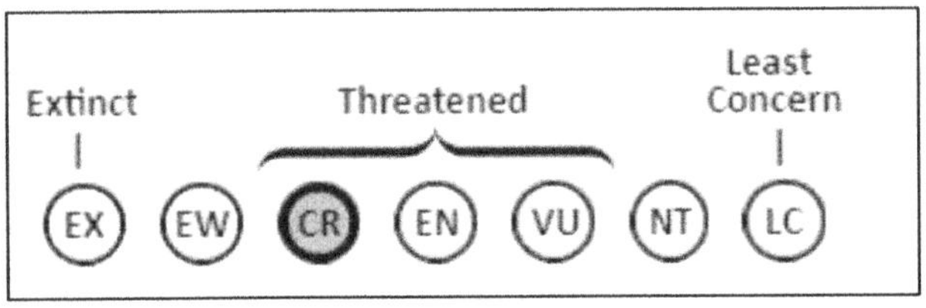

[1]

NCA = CR = Critically endangered

EPBC = CE = Critically endangered

To learn more about these birds, watch the video from Taronga Western Plains Zoo: https://www.youtube.com/watch?v=Y4680qGgUrE

Australian pelican

Pelecanus conspicillatus

- Australian pelicans have an extremely light skeleton, which weighs only 10% of the whole weight of their body. They are one of the largest flying birds in the world (1.6 - 1.8 m long, 2.3 - 2.5 m wingspan), only second to the snowy albatross, and can stay in the air for over 24 hours, travelling hundreds of miles as they soar in the breezes at heights of up to 3km and reaching speeds up to 56km per hour!
- The Australian pelican's beak is the largest of all birds, growing up to 50cm long—a world record for the longest bill of any bird! It is so large that it can hold from 9 to 13L of water. There is a hook at the end of the bill, which they use to hold onto slippery fish. They can eat around 9kg of food a day.
- Pelicans are mouth-breathing birds. To cool off in hot weather, they pant like dogs to circulate air over their **saturated** pouch, which serves as a heat exchanger.
- They live in **colonies** with thousands of other pelicans, and often work together to catch fish and protect their young. The young leave the nest at around 28 days and form **creches** of about 100 chicks, which become independent at around the age of 4 months when their parents stop feeding them.
- A group of pelicans is called a brief, squadron, or pod.
- The pelican family is at least 30 million years old!

"A wonderful bird is the pelican...
His beak can hold more than his belly can.
He can hold in his beak
Enough food for a week!
But I'll be darned if I know how the hellican?"

Dixon Lanier Merritt

Pelicans DO NOT store food in the pouch of their bills!

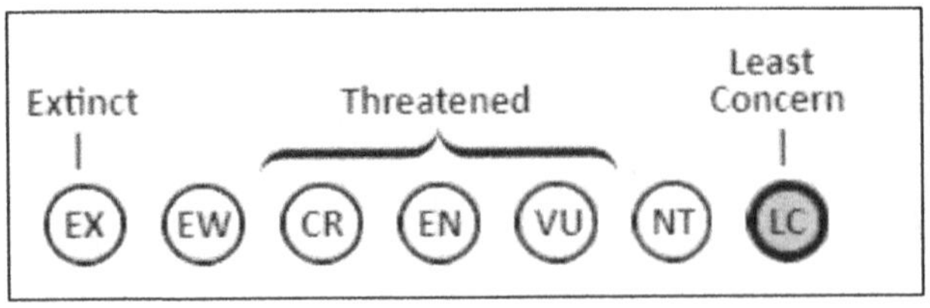

[1]

Not categorised on either NCA or EPBC lists.

*Free Printable Pelican Colouring Page – Simple Mom Project. https://simplemomproject.com/pelican-coloring-page/

Koala

Phascolarctos cinereus

NATIVE TO AUSTRALIA

- The word "Koala" is thought to originate from one of the Australian Aboriginal languages and means 'no drink'. Koalas drinking water is uncommon although they may do during heat waves, droughts or bush fires.
- The average koala weighs about 9-10kg. They have five **digits** on each paw, but on the front paws, two are **opposing**, like thumbs, to help them grip trees and branches better.
- Koalas and humans have almost identical fingerprints!
- They have a specialist diet of gum leaves, poisonous to most other animals. They can eat up to 1kg of leaves a day. Koalas have an excellent sense of smell and will sometimes sniff leaves before tasting them, favouring the more **nutrient-rich** leaves.
- They can sleep for 18-20 hours a day. **Digesting** all those leaves takes a long time, and they do not get a lot of energy from just eating leaves.
- They are the only surviving members of their family; their closest living relatives are wombats.
- Koalas are **marsupials**. They don't live in **social** groups.
- Did you know? Female koalas are only pregnant for around 35 days. <u>In comparison, an elephant is pregnant for approximately 620 days!</u>[13]
- Baby koalas are called **joeys**. They are born without any hair, blind and without ears, and are about the size of a jellybean.
- When ready to leave the pouch, a **joey** will hitch on mum's back for six months to learn **critical survival** skills such as **foraging** for food and avoiding predators. [13]
- Koalas are usually slow-moving but can gallop up to 30km/hr when threatened.
- The "Black Summer" bushfires in 2019-2020 are thought to have been responsible for the deaths of 80% of koala populations in some regions. During bushfires, koalas unable to escape burning areas can be killed directly by flames, their **habitats** are destroyed, there is reduced food availability and an increased risk of **predation**. Such **traumatic** events also impact stress levels and **disease susceptibility**, disrupt breeding patterns, and cause **fragmentation** of **populations, hindering genetic diversity**.

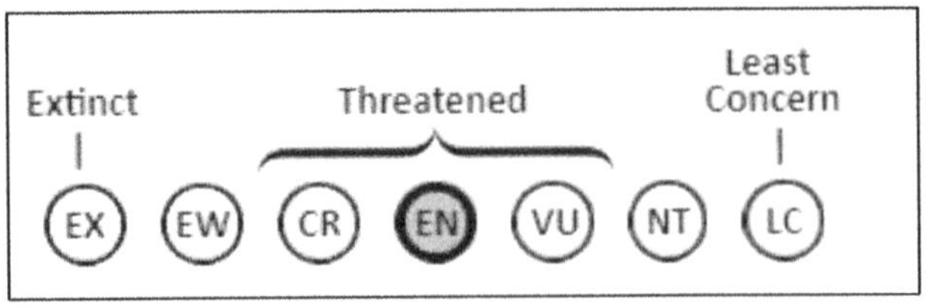

[1]

NCA List = E = Endangered

EPBC List = E = Endangered

Watch this short video from ABC iview on YouTube:
The beautiful bond between koala and eucalyptus tree | Nature's Great Divide
https://www.youtube.com/watch?v=QC2V311zieo

Long-nosed potoroo

Potorous tridactylus tridactylus

NATIVE TO AUSTRALIA AND TASMANIA

- Potoroos are **marsupials** and are mainly **nocturnal**. They hide during the day in thick **vegetation** in forested areas and coastal **heaths**, but in the winter months, they might occasionally be out searching for food during the day.
- Potoroos prefer wet forests and wet scrublands with thick understorey **vegetation**, essential for cover. Eucalypt forests are important because potoroos rely on **fungi** associated with these trees.
- Potoroos are small members of the same family as bettongs and desert rat-kangaroos, looking a little like a bandicoot until they hop away with their front feet tucked tight to their chests.
- Five **species** in the Potoroidae family are native to Victoria, three bettong **species**, and two potoroo **species**; sadly, all bettong **species** are **locally extinct**.
- Long-nosed potoroos are **solitary** in the wild although when the females have young to care for this is not so. They are not **territoria**l.
- The female potoroo carries nesting material with her tail, which is **semi-prehensile**.
- Their fur is brown-grey with **rufous** tinge on flanks and pale grey underparts.
- They grow to have a head and body measurement of around 340 to 400mm and weigh 660 to 1600g. Their hind foot length is shorter than their head length.
- They can live up to 12 years in captivity but, in the wild average 5 to 6 years.
- **Fungi** form a large part of the potoroos' diet in the wild, and it is thought that they may play an important role in spreading environmentally important **fungi** within their **ranges**. Their diet also includes seeds, fruits, **vegetation**, **tubers**, and insects in the soil.
- Long-nosed potoroos give birth to one single youngster weighing 0.3g. The youngster stays in the pouch for between 120 and 130 days, stays near the mother after leaving the pouch, and still feeds from her for about another 40 days before becoming independent.
- Threats include the loss and **fragmentation** of natural **habitats**, which results in the **isolation** of **populations**, and **predation** by introduced cats and foxes.

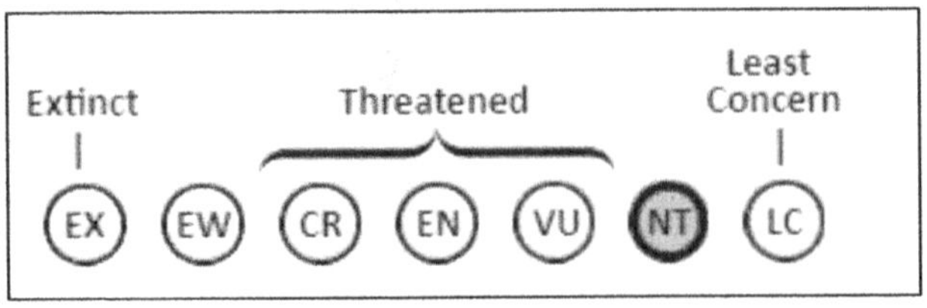

[1]

NCA List = V = Vulnerable

EPBC List = V = Vulnerable

Sugar glider

Petaurus breviceps

NATIVE TO AUSTRALIA

- **Nocturnal, arboreal, omnivorous**, small gliding possum. It can glide 50m or more, although it loses about 1m for every 1.82m in height.
- Latin name translates as "short-headed rope-dancer," referring to their **acrobatics** in the **tree canopy**!
- The oldest fossils in Australia date back to 4.46 million years.
- They grow up to 24 to 30cm from the tip of the tail to the nose, with females weighing up to 115g and males up to 140g.
- The sugar glider has a squirrel-shaped body.
- Males have bald spots on their chests and heads where they have scent glands, which are used to mark their territory and group members; females do not have scent glands in these places on their bodies, nor do they have the bald spots, so that is useful in identifying whether the animal is male or female!
- The fur coat is thick and soft, usually blue-grey, although **variations** are known to be yellow, tan, and even **albino**. The gliding **membranes** are called **patagia**.
- They have a long grey tail, which is weakly **prehensile**.
- Each foot has five **digits**, with an **opposable** toe on each hind foot. These **opposable** toes do not have claws and bend like our thumbs do so that they can touch all the other **digits** allowing the glider to grasp branches firmly. The second and third **digits** of the hind foot are partially fused together. The fourth **digit** of the forefoot is sharp and elongated, which enables the glider to dig out insects from underneath the bark of trees!

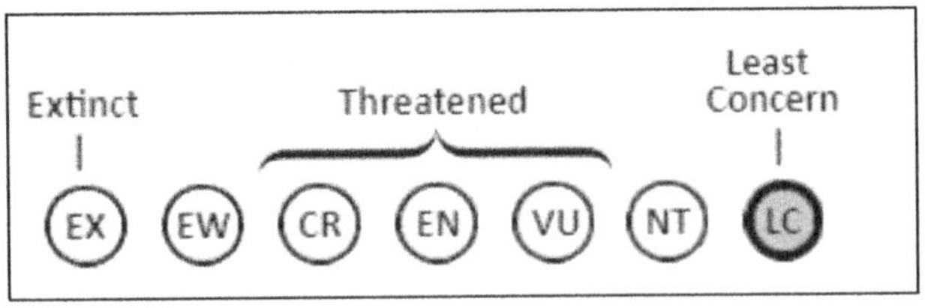

[1]

NCA = C = Least concern

EPBC lists

Powerful owl

Ninox strenua

- The powerful owl is the largest owl in Australia. Males are larger and stronger than females.
- Powerful owls are found in coastal forests in eastern Australia, usually not more than 200km inland, with a large home **range** of between 800 to 1000ha (although t**erritories** can be larger when food is scarce). They are **raptors** and are typically **territorial**.
- They like to live in tall and **humid** forests where their **prey** also lives.
- As an **apex predator**, the powerful owl mostly hunts **mammals** living in trees, in particular small to medium-sized **marsupials**. This **prey** provides about three-quarters of the powerful owl's diet.
- Like other owls, powerful owls have **evolved** to have some **unique adaptations.** They can hunt in almost total darkness due to the **unique anatomy** of their eyes, which have lots of light-sensitive cells. They also have remarkable hearing, which allows their brain to build almost a 3D map of space to locate their prey.
- Owls cannot move their eyes up or down or from side to side? This is because their eyes are **cylindrical** and fixed in place, not like human eyes at all! No wonder they always look as if they are looking straight at you! They are!!
- Owls have three eyelids: one for blinking, one for sleeping, and one for keeping their eye structures clean.
- Owls, also, have 14 **vertebrae** in their necks, allowing them up to 270 degrees of movement, meaning that they have a significant range of motion in their necks, and can look behind themselves or up and down! We humans only have seven **vertebrae** in our necks, so we have significantly less range of movement, up to a maximum of 180 degrees.. Owls can achieve half again of what we humans can in terms of neck rotation!
- Owls also have poor day vision - they do not have as many 'rods' in their **retinae** but, despite poor day vision, their night vision is similar to what we would see using night vision goggles!
- There can be some flexibility in terms of the **habitats** of powerful owls when they are not nesting.
- They have long intervals between the laying of eggs and the hatching of their **clutches**.
- The only continent in the world where owls are not found is Antarctica.

[1]

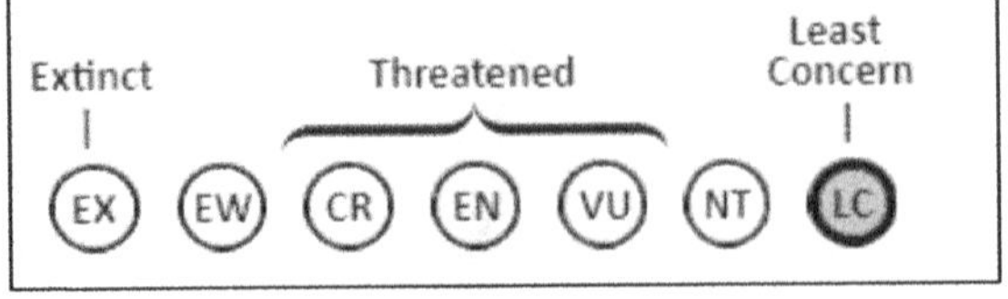

EPBC List = Not listed

Eastern bandy bandy

Vermicella annulata

NATIVE TO AUSTRALIA

- The bandy bandy is also commonly known as the hoop snake. There we are! I thought hoop snakes were an Australian **myth**!
- It is a smooth-scaled and **glossy** snake with black and white rings which go right around the body.
- It is weakly **venomous**, and a bite results in **localised** symptoms. It is generally considered harmless, with a small mouth and a non-aggressive nature.
- Male bandy bandy snakes are much smaller than females with an average length of 40cm compared to around 60cm for females.
- Bandy bandy snakes feed almost exclusively on other snakes, mostly blind snakes. They find their **prey** by flicking their tongues, searching for **chemical cues** left behind by blind snakes.
- The eastern bandy bandy is found in a range of **habitats**, including acacia, mulga, and mallee scrubs, wet coastal forests, **spinifex** desert sandhills, and **savannah** woodland.
- They are **nocturnal**, burrowing snakes, found under stumps, rocks, logs and beneath the surface of the soil.
- **Predators** include owls, feral cats, foxes, and larger **reptiles**. The bandy bandy uses its striking colour pattern to confuse potential **predators** by moving fast in dim light conditions. It also takes a **defensive stance**, twisting its body into multiple loops, reaching a height of 15cm with head and tail remaining at ground level. This behaviour combined with its colour and pattern of markings can leave a potential **predator intimidated** and confused.

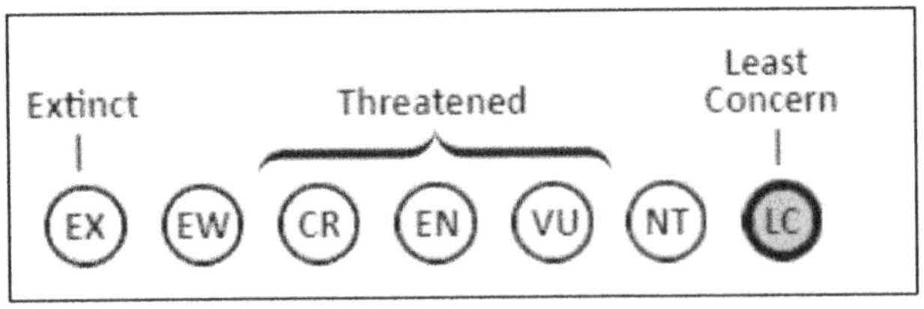

[1]

NCA List = Not listed (Qld)

EPBC List = Not listed

Australasian bittern

Botaurus poiciloptilus

- The Australasian bittern, is a large bird in the heron family nicknamed the "**bunyip** bird”.
- It is **secretive** and has a **distinctive** booming call, so it is more often heard than seen.
- It feeds on **frogs, eels**, freshwater **crustaceans,** and other **aquatic** animals.
- It is a **solitary** nester on the ground in **dense** wetland **vegetation** on **trampled** reeds and other plants.
- These birds are hard to find, and so **monitoring** of this **species** mainly relies on the ability to count the male birds based on their notable breeding calls. The best time to detect Australasian bitterns has been shown to be one hour before sunrise, in September (**Austral spring**), on a moonlit night with no clouds or rain.

- The **bunyip** is a **creature** from the **aboriginal mythology** of southeastern Australia, said to lurk in **swamps, billabongs**, **creeks,** riverbeds, and **waterholes.**

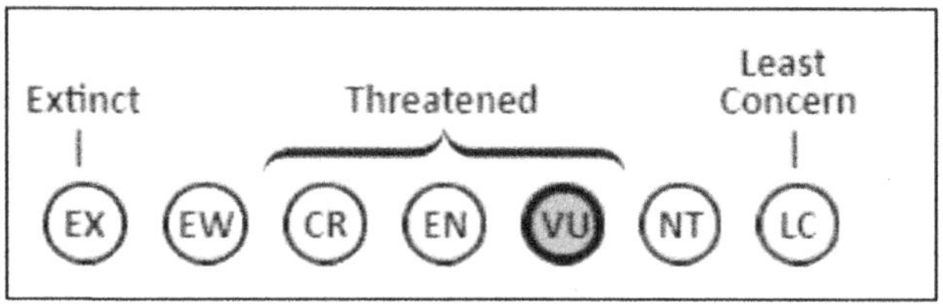

[1]

NCA List = E = Endangered

EPBC List = E = Endangered

Moonlight jewel

Hypochrysops delicia

FOUND IN AUSTRALIA WITH ONE SUBSPECIES IN THE MALUKU ISLANDS IN INDONESIA

- What a beautiful name! The moonlight jewel!
- The wingspan of this species is about 40mm.
- Take out your ruler and measure what is 40mm to see how small it is!
- It is primarily found in South Australia but may also be seen in the Wallum from time to time. Interestingly, it also has a similar relationship with ants, members of the *CREMOGASTER* species, that the Illidge's ant-blue does!
- The moonlight jewel is a member of the *Lycaenidae* family. It is found in eastern Australia.
- The **larvae** feed on *Acacia* species.

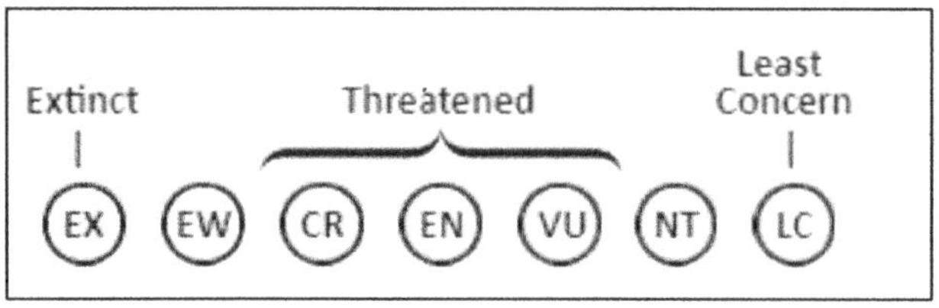

[1]

EPBC List = Not listed

Not listed on IUCN Red List

Australian shelduck

Tadorna tadornoides

- The Australian shelduck, also known as the chestnut-breasted shelduck or mountain duck, is a large, distinctive waterfowl species found in southern Australia.
- They inhabit freshwater and some saltwater **habitats** near freshwater sources, and are often seen in pairs or large **flocks**.
- They are among the largest duck species in Australia.
- Shelducks are a group of large goose-like ducks , part of the bird family *Anatidae*. The genus name *Tadorna* comes from Celtic roots and means "pied waterfowl".
- They are known for their striking **plumage**, including a chestnut breast and black body with white markings. Males have a black head, females have white around the eye and **bill**, a white neck ring, and a chestnut breast.
- Australian Shelducks stand upright when grazing on land and also sit high in the water.
- Males produce a deep, harsh honking sound, while females make a high-pitched "ank-ank" sound.
- They nest in tree hollows, holes in banks, or even rabbit burrows, often lined with down.
- Chicks leave the nest within a couple of days, led by their parents, and often form **crèches** with other families.
- Australian Shelducks are **omnivores**, feeding by grazing, dabbling and filtering. They eat **tubers** and other plant material, such as saltbush and samphire, along with a range of insects, **crustaceans** and molluscs.
- They have adapted to various **habitats**, including coastal areas, freshwater swamps, and even inland lakes and gravel pits.
- After breeding, they may **migrate** to wetlands in large **flocks** of 1000 or more to **moult**
- their feathers.
- They are generally **monogamous**, with both parents participating in raising the young.
- While not a primary game bird, they may be hunted occasionally due to their habit of grazing on crops.

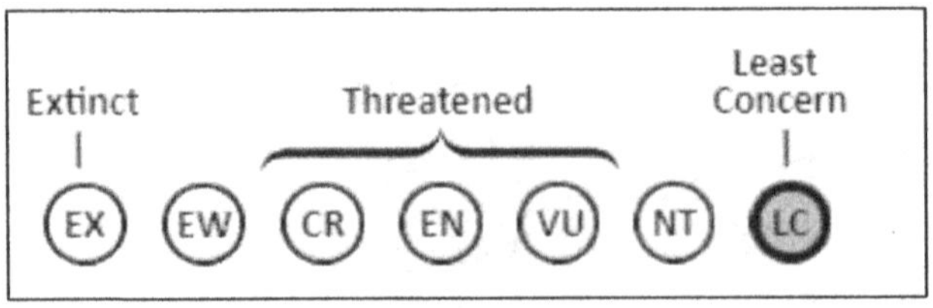

[1]

EPBC List = Least concern

Burrowing bettong (Boodie)

Bettongia lesueur

- A bettong is about the size of a rabbit, with body length ranging from 30-38cm among species. All bettongs have long tails, roughly equal to body length. Body weight ranges from 1.2kg in the smallest species (northern bettong) to 2.8kg in the rufous bettong, the largest of the group. Males tend to be slightly larger than females. Bettongs have furry coats, ranging in colour from grey to ginger and brown.[14]
- There are five bettong species: Eastern (or Tasmanian) bettong (*Bettongia gaimardi*), burrowing bettong (or Boodie) (*B. lesueur*), brush-tailed Bettong (or Woylie) (*B. penicillata*), northern bettong (*B. tropica*), and rufous bettong (*Aepyprymnus rufescens*).
- The **burrowing** betong or Boodie, also known as Lesueur's rat-kangaroo, is a small, furry, rat-like **mammal native** to Australia.
- It was declared a pest in the 1900s and was wiped out by the 1960s; however, the loss of the Boodie and other ground-**foraging** animals has **degraded** soil quality. Populations persisted on three west coast islands (Bernier**,** Dorre**,** and Barrow), and the Boodie has been reintroduced to the mainland since the 2000s.[15]
- The word "Boodie" comes from its name *boodi* or *burdi* in Noongar, an Aboriginal language spoken in southwestern Australia.[15]
- Boodies are fascinating **marsupials** known for their unique **burrowing** habits, which are unusual for **macropods**. They create **complex burrow** systems, some with over 100 entrances, that can house many individuals and provide shelter from the heat and **predators**.
- They are **nocturnal**, meaning they are active at night, and tend to live in groups within their warrens.
- Their digging activities improve soil health by increasing water infiltration, soil turnover, and seed dispersal.
- They **forage** for a variety of foods, including roots, **tubers**, seeds, fruits, and **invertebrates**.
- Burrowing bettongs can breed throughout the year, with females capable of having multiple **litters** per year.
- Once common, they are now considered vulnerable due to threats like habitat loss and **predation** by introduced **species** like foxes and cats.
- They are sometimes called "rat-kangaroos" due to their size and the combination of hopping like a kangaroo and having a rat-like appearance.
- European rabbits competing for resources and **habitat** loss have also impacted their populations.

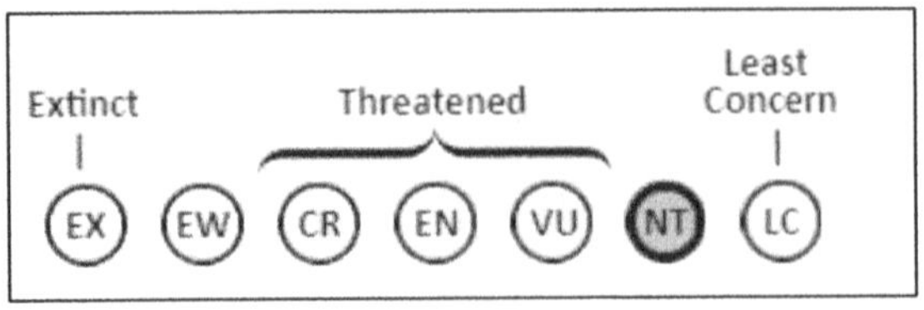

[1]

EPBC List = V = Vulnerable

Brown goshawk

Accipiter fasciatus

- The brown goshawk is a widely distributed **raptor** in Australia, known for its adaptability and hunting prowess. It is a medium-sized hawk, with a brown head, greyish-brown wings and back, and a reddish-brown belly with white bars.
- When comparing the brown goshawk to the red goshawk, in the case of the red goshawk, the wings are longer and more pointed, and the tail is shorter. The red goshawk occurs across much of northern Australia, from near Broome in the south-west Kimberley to south-eastern Queensland.
- They are frequently spotted in forested areas, including suburban areas with nearby trees, and even towns and cities.
- Brown goshawks are versatile **predators**, using various hunting techniques like stalking, pouncing from concealed perches, and chasing prey in flight. They are known to hunt a variety of prey, including birds, reptiles, insects, and even introduced rabbits. They are known to take **carrion** as well.
- They often hunt from a low, concealed perch, using sudden bursts of speed to strike their **prey**. They then take the **prey** to a perch to partially pluck it before eating.
- Brown goshawks build large stick nests in tall trees, often near water or forest edges. They may reuse old **raptor** nests and line them with fresh eucalypt leaves.
- Both parents **aggressively** defend their nest and territory. Established pairs often reuse the same nesting area annually.
- Brown goshawks are **sexually dimorphic** - female brown goshawks are considerably larger than males.
- They have a distinctive call described as a loud, rising "keek-keek-keek" or a slow, drawn-out "ee-you-wick, ee-you-wick", with females having a deeper voice than males.
- They are found throughout Australia, including Tasmania, and also on islands like New Guinea and New Caledonia.
- Brown goshawks have **adapted** well to suburban environments and are frequently seen in home gardens.
- They are currently classified as least concern meaning their populations are stable, though they may face threats from habitat loss and **rodenticides**.

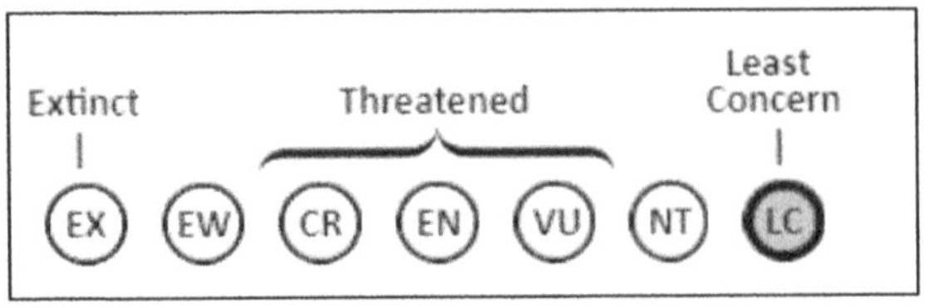

[1]

EPBC List = Least concern

Eastern quoll

Dasyurus viverrinus

- The eastern quoll (formerly known as the eastern native cat) is a medium-sized **carnivorous marsupia**l, and one of six extant **species** of quolls. **Endemic** to Australia, they occur on the island state of Tasmania, but were considered **extinct** on the mainland after 1963.[16] The **species** was reintroduced to fox-proof fenced sanctuaries in Victoria in 2003 and to the Australian Capital Territory in 2016.
- Eastern quolls are known for their spotted coats and are fascinating marsupials. They are **nocturnal, solitary** hunters that **prey** on insects, small mammals, and birds.
- Their numbers on mainland Australia declined due to introduced **predators**.
- Two colour morphs exist, and eastern quolls can be either black with white spots or fawn with white spots, which is an evolutionary adaptation to their environment.
- They are primarily active at night, hunting for food, and prefer to be alone, scent-marking their territories.
- While primarily **terrestrial**, they are also skilled climbers, using their **agility** to hunt and find shelter.
- Eastern quolls are listed as endangered on the IUCN Red List, with the main threats being **habitat** destruction and introduced **predators** like foxes and cats.
- Like other **marsupials**, female eastern quolls have a pouch where their young, called **joeys**, develop.
- While they primarily eat insects, they are **opportunistic carnivores**, also consuming small mammals, birds, and reptiles, and will scavenge on **carrion**.
- They can be distinguished from all other species of quoll by the presence of only four toes, rather than five, on the hind feet, lacking the hallux.[17]
- Eastern quolls use various **dens** for shelter, including underground burrows, hollow logs, and rock **crevices**, often alternating between them.
- They communicate and mark their **territories** using scent marking, though their scats are not placed at specific latrines.

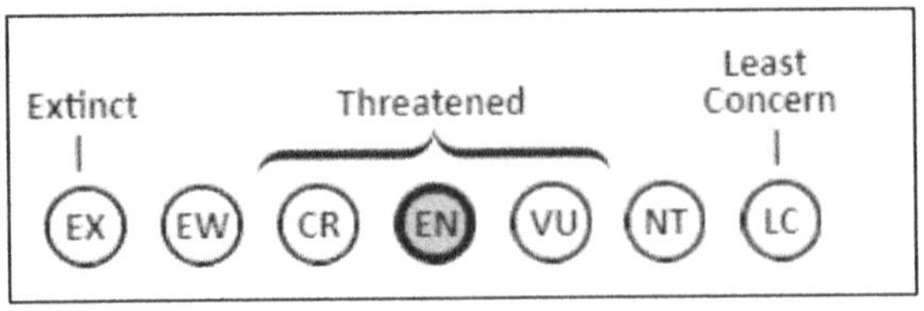

[1]

EPBC List = E = Endangered

Orchard swallowtail

Papilio aegeus

- Orchard swallowtail butterflies are large and striking insects known for their black and white patterns. They are a common sight in gardens and citrus orchards due to their caterpillars feeding on citrus leaves. Interesting facts include their caterpillar's ability to mimic bird droppings for camouflage, and their **osmeterium.**
- Orchard Swallowtails are one of the largest butterflies in Australia, usually having a wingspan between 10-14cm.
- Male butterflies are black with creamy-white spots and a red spot on their hindwings, while females are browner with a white patch.
- Young caterpillars are green and brown with white markings and spines, and they look like bird droppings, which helps them hide from predators.
- Mature caterpillars have a red, Y-shaped organ (**osmeterium**) that they can push out when they feel threatened. It emits a foul, rotting citrus smell to deter **predators**.
- They are commonly found in Queensland and along the east coast of Australia, particularly in areas with citrus trees.
- Caterpillars primarily eat the leaves of citrus trees like oranges, lemons, and limes, which is why they are often seen in gardens and orchards.
- They have a fast, direct flapping flight, especially males, who can be **territorial** and chase away anything black and white that enters their area, even magpies!
- The caterpillar goes through five **instar** stages before emerging as a beautiful butterfly.
- While they are beautiful, they are sometimes considered a pest by citrus farmers because of their caterpillars' appetite for citrus leaves.

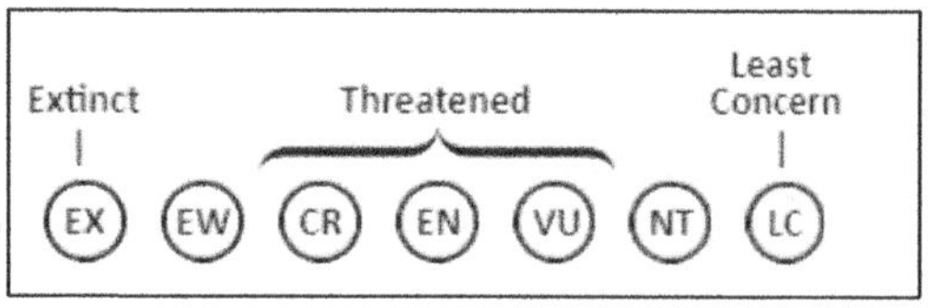

[1]

EPBC List = Not listed

Red-tailed tropicbird

Phaethon rubricauda

- The red-tailed tropicbird is the rarest of the tropicbirds and is native to the tropical parts of the Indian and Pacific Oceans. It is similar to a tern in appearance, with almost all-white plumage, a black mask, and a red bill.
- The bird has a **streamlined** but solid body shape, a wingspan of 111-119cm, a weight of 800g, and a length between 95-104cm, including the 35cm long red tail streamers in adult birds. These **tail streamers** are highly prized by the **Māori people** in the northernmost regions of New Zealand. Still, the birds are seen more frequently in ocean areas with water temperatures ranging from 23 to 30°C.
- Common names include red-tailed bos'nbird or silver bos'nbird and strawtail. The New Zealand Māori call it amokura, and the native Hawaiians koa'e 'ula.
- The red-tailed tropicbird eats mainly squid and flying fish, which it catches by plunge-diving into the ocean or sometimes catching flying fish in the air.
- The red-tailed tropicbird nests in loose **colonies** on **oceanic** islands; the nest itself can be a scrape found on the face of cliffs, on a sandy beach or in a **crevice**.
- Only a single egg is laid and is **incubated** by both sexes for about six weeks. The parents make long trips **foraging** for food during the **incubation** period, sometimes for as much as six days.
- Threats include **habitat** destruction, pest introduction, and **predators** such as rats and **feral** cats that **prey** on eggs and young at nesting sites.
- Red-tailed tropicbirds are generally silent while flying. Birds may greet their mates with a squawk when leaving or arriving at the nest. The chicks chatter repetitively, as a begging call, whenever the parents are near, and the adult birds give a low, growling defence call.
- Strong winds can occasionally blow them inland, which explains sightings that have been recorded away from the coastlines and their preferred **habitats**.
- The red-tailed tropicbird is strong in flight but walks with a shuffle on land with some difficulty, landing by flying into the wind, then **stalling** and dropping to the ground. Nests are usually no more than a metre from the edge of the shaded area or scrub so that it does not have to walk too far.
- They can hover in midair by flying into the wind; pairs may even fly backward and circle each other during courtship. They can be **territorial,** with bill-jabbing and fights breaking out, the two **combatants** locking bills and wrestling for up to 90 minutes.

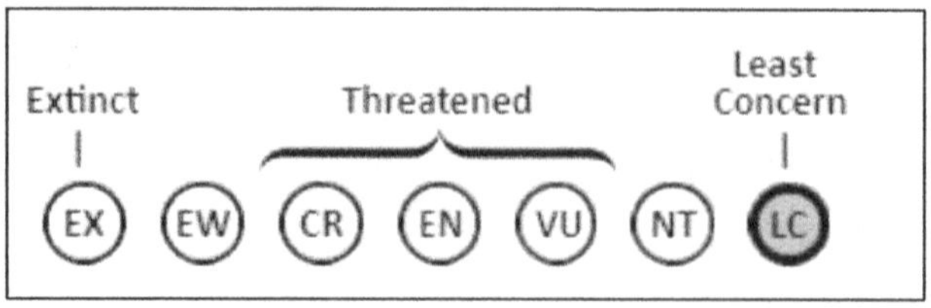

[1]

EPBC List = Not listed

Sooty albatross

Phoebetria fusca

- Sooty albatrosses are seabirds with a striking appearance and interesting behaviours.
- The sooty albatross is also known to sailors as the Quaker,
- They have a dark, sooty-brown to black plumage, a white ring around their eye, and a black beak with a yellow or orange stripe. T
- These birds are known for their impressive wingspan, which allows them to glide effortlessly over the ocean for long distances.
- Sooty albatrosses are medium-sized albatrosses, with a wingspan of about 2m.
- Their plumage is a dark, sooty-brown to black colour. The head and sometimes the tail can be a darker shade than the rest of the body. They have a distinctive white ring around their eye and a yellow or orange stripe on their beak.
- In general, the sooty albatross can be found **foraging** over deep ocean. Non-breeding adults do not often venture out of their native oceans but have a broader **foraging** range, while breeding adults **forage** closer to their breeding sites (within 600km) and **juvenile** birds tend to be confined to **subtropical** waters.
- They eat fish, **crustaceans**, squid, and **carrion**. They often **forage** alone, but may join mixed-species **flocks** at times, especially when following fishing boats.
- Sooty albatrosses are masters of the wind. They use their long wingspan to glide for long distances, often without flapping their wings, making them very efficient flyers.
- They breed in loose **colonies** on cliffs or steep slopes. A single egg is laid in October or November, and both parents help to incubate it and raise the chick.
- Like all albatrosses, sooty albatrosses drink saltwater. They have special glands above their eyes that help them get rid of the excess salt, which can sometimes drip from their nostrils.
- Sooty albatrosses can live for a long time, with some individuals living for over 60 years.
- Sooty albatrosses are currently listed as near threatened due to threats like bycatch in fishing gear and climate change.

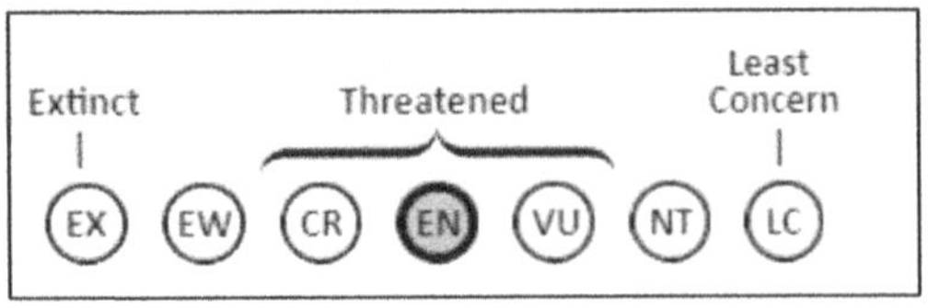

[1]

EPBC List = V =Vulnerable

Yellow-bellied glider (south eastern subspecies)

Petaurus australis australis

- Yellow-bellied gliders are fascinating **marsupials**. They have a black stripe down their back and tail, and their belly fur can range from white to yellow, with older gliders having more yellow fur.
- They are **social** creatures, living in family groups and defending large **territories**. Their diet includes **nectar**, sap, and insects.
- They are gliding experts having a special **membrane** of skin called a **patagium** between their wrists and ankles that stretches out, allowing them to glide through the air like a flying squirrel. They can glide up to 100m in one jump, and some have been recorded gliding even further.
- They are highly **adapted** to life in the trees, spending almost all of their time in the forest **canopy**.
- They create V-shaped incisions in the bark of eucalypt trees to access sap, a key part of their diet. Yellow-bellied gliders are highly selective about which trees they feed on and may revisit the same trees for over 20 years. Their diet includes tree sap, **nectar, honeydew, pollen**, and insects.
- They live in family groups, with males, females, and their young, and these groups defend large **territorie**s.
- Yellow-bellied gliders are nocturnal and vocalise with loud shrieks and calls, especially at night, that can be heard up to 500m away. Their calls serve both communication within family groups and **territorial** defence.
- They rely heavily on eucalypt trees for sap and also utilise hollows in these trees for denning and breeding. They require hollows in trees for shelter and breeding, making the preservation of old-growth forests crucial.
- **Habitat loss** and **fragmentation** due to deforestation and timber harvesting are major threats to their survival. Populations in the south-east of Australia are considered **vulnerable**, with a significant population decrease reported in recent years.

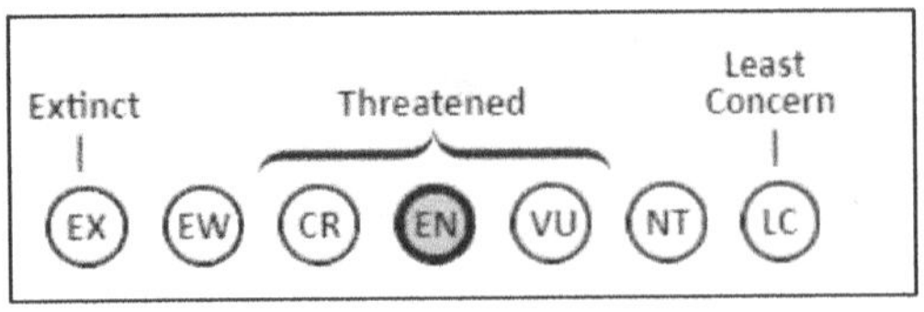

[1]

EPBC List = V =Vulnerable

Black-tailed godwit

Limosa limosa

- There are four recognised **subspecies** of the black-tailed godwit, all with orange head, neck and chest in breeding **plumage** and dull grey-brown winter colouration, and a distinctive black and white wingbar at all times.
- Known for their long **migrations**, these elegant wading birds breed in wet grasslands and bogs in places like Iceland and Siberia, but they spend their winters in warmer areas like southern Europe, Africa, and Asia, **migrating** in **flocks**.
- Black-tailed godwits are tall birds with long legs that are great for wading in shallow water and a long, slender beak 7.5-12cm long, that they use to **probe** the mud for tasty insects, worms, and small creatures.
- In flight, its bold black and white wings and white rump can be seen readily. When on the ground it can be difficult to separate from the similar bar-tailed godwit, but the black-tailed godwit's longer, straighter **bill** and longer legs are diagnostic. Black-tailed godwits are similar in body size and shape to bar-taileds, but stand taller.[19]
- The most common call is a strident *weeka weeka weeka*.
- They are amazing travellers! They fly long distances from their breeding grounds to their wintering areas, often in large groups.
- These birds are very loyal. They often form strong bonds with their mates that can last for many years, coming back to the same breeding grounds together each spring to raise their chicks together. Both the male and female black-tailed godwit work together to incubate the eggs and take care of the chicks after they hatch.
- A study of the Icelandic population showed that despite spending winter apart, pairs are reunited on their breeding grounds within an average of three days of each other. If one partner does not arrive on time, 'divorce' occurs. They nest in loose **colonies**. Unpaired males defend a temporary **territory** and perform display flights to attract a mate. Several nest scrapes are made away from the **courtship territory** and are defended from other godwits.[20]
- The species breeds in fens, lake edges, damp meadows, moorlands and bogs and uses estuaries, swamps and floods in (the northern hemisphere) winter; it is more likely to be found inland and on freshwater than the similar bar-tailed godwit.

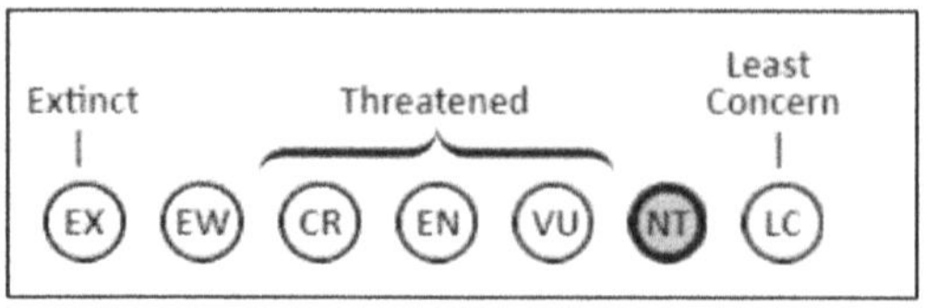

[1]

EPBC List = E =Endangered

Red knot

Calidris canutus

- Red knots are shorebirds known for their incredible long-distance **migrations**, travelling thousands of miles between their Arctic breeding grounds and wintering sites in places like Australia and South America.
- There are six **subspecies** of red knots with the Eurasian populations wintering in Africa, Papua New Guinea, Australia and New Zealand.
  Red knots are among the longest-distance **migrants** in the animal kingdom, flying over 14,000km each way between their breeding and non-breeding areas.
- Despite being small birds with a wingspan of only about 47-53cm, they can fly non-stop for very long distances.
- Their diet varies according to season. On the breeding grounds, knots eat mostly spiders, **arthropods**, **larvae**, and some plant material obtained by surface pecking. Red knots love to eat **bivalves** like clams and mussels, which they find by **probing** the mud and sand at low tide. They eat the shells whole, and their strong gizzards crush them to digest. They also eat worms and **crustaceans**.
- This species forms enormous **flocks** when not breeding. They often gather in large, tight groups, especially on sandy beaches and mudflats, and can even intermingle with other shorebirds like great knots.
- Red knots change colour! Their feathers turn a rusty red when they are heading north to breed and then return to a more greyish colour for their wintering journey.
- They breed in cold climes, including the Arctic tundra, Greenland, the Taymyr Peninsula, the Chukchi Peninsula and some Siberian islands, building shallow nests on the open ground with their eggs **camouflaged** with blotches and spots.
- During their long **migrations**, and in the case of the North American population, places like the Delaware Bay are vital rest stops for them, where they feed on horseshoe crab eggs to get ready for the final leg of their journey.
- Red knots are facing challenges due to climate change, which could affect their Arctic breeding grounds and the availability of important food sources.

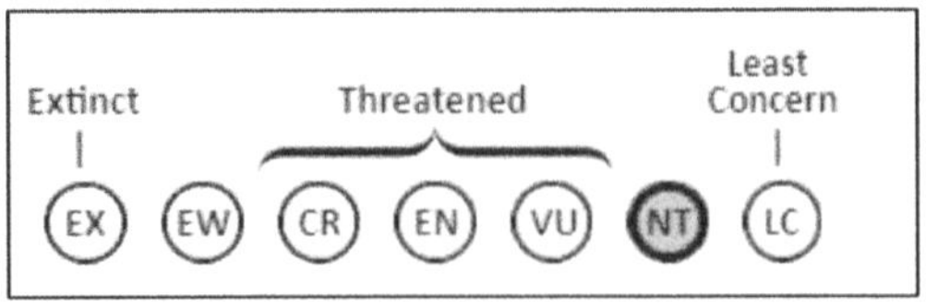

[1]

EPBC List = E =Endangered

Macleay's swallowtail

Graphium macleayanum

- Macleay's Swallowtail is a beautiful butterfly species identified by the tails on its hind wings and the green colour on the undersides of its wings. It also has green legs that match its wings, an unusual feature of this butterfly.[21]
- Most of the butterflies in this family are large in size, ranging in size from 5.3cm to 6cm, with wingspan up to 8cm, and with brilliant colours. They are called swallowtails because some of the species have tailed hindwings. [21]
- These stunning butterflies often feed on flowers with their wings rapidly vibrating.
- The Macleay's swallowtail is native to the wetter coastal and mountain parts of eastern Australia, from northern Queensland south to Victoria and Tasmania. [21] They also occur on Lord Howe Island and in the highlands of Papua New Guinea.
- They live in **urban** areas, forests and woodlands, and heath.
- They are commonly seen in urban Sydney between August and April.
- The caterpillars of Macleay's swallowtail have adapted to feed on a variety of plants, including Sassafras leaves and the introduced Camphor Laurel (*Cinnamomum camphora*).
- The caterpillars go through several moults during their development; initially, they have a black hump and black forked tail. Then they become plain green with small white dots over the body and two narrow yellow lines along the back. At these stages, they are called **instars**.
- The caterpillars grow to around about 4cm in length.
- The eggs are round and pale green. They are laid singly on young shoots of a food plant.
- The caterpillar is green with a humped **thorax**. Initially, it has a black hump and a black forked tail. Later, it becomes plain green with small white dots over the body, and two narrow yellow lines along the back.
- Males **congregate** around hilltops where they can be seen defending their **territory** from rival males and **courting** passing females.
- They can usually be seen flying above the eucalypts at the highest point of a hill, rarely coming down to a catchable height.
- They are related to other members of the butterfly family *Papilionidae* such as the Orchard Butterfly, Canberra's biggest, and the magnificent blue Ulysses Swallowtail of north Queensland.

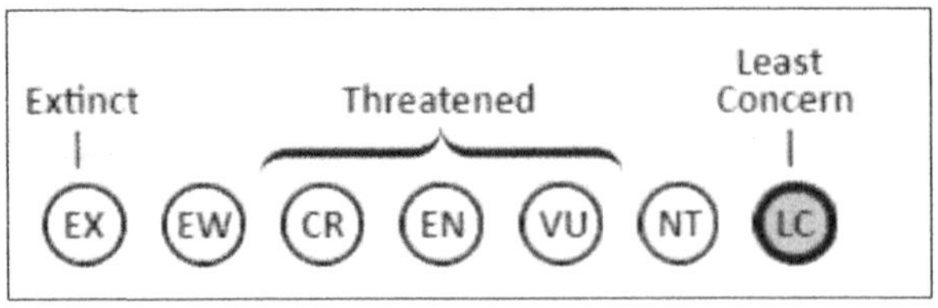

[1]

EPBC List = Not listed

Eastern curlew

Numenius madagascariensis

- Eastern curlews, recognised by their "cuuuurlew" call, are the largest shorebirds in Australia, reaching heights of up to 60cm.
- They feature an impressive, **distinctive** curved beak averaging 185mm in length which is used to probe the mud and did up crabs, their main food source in Australia. Eastern curlews also eat molluscs, and worms. During their migration, they even eat berries.
- It is a bulky, dark-streaked brown wader, with a long neck and legs.
- When flying, the barred flight feathers are visible, lighter under the wings and dark above.[22] It has a **plumage** of mottled brown and tan feathers. Average size is 63cm and the average weight is 900 grams.[22]
- Their wing beats are slow and deliberate, unlike the rapid beats of the Whimbrel.[22]
- In Australia, the Eastern curlew is classified as critically endangered, with their **global population** declining by approximately 80% over the past three decades.
- These **migratory** wading birds depend on **intertidal** mudflats and sandflats, often with beds of seagrass, on sheltered coasts, especially estuaries, mangrove swamps, bays, harbours and lagoons, for food and shelter.
- However, the development and erosion of these mudflats along their **migratory** paths pose a significant threat to their survival.
- Eastern Curlews are quite shy and easily disturbed, so it's important to give them space and keep pets on leashes when you're near their feeding or roosting areas, and avoid using drones nearby. They are wary birds, quick to take flight.
- The eastern curlew begins its journey from its breeding grounds in Russia and northeastern China, travelling over 10,000km to reach warmer climates in southern feeding areas, such as the coastal regions around Cleveland in southern Moreton Bay, which remain one of their last **refuges** on Australia's eastern coast. They can also be found in the Coorong region of South Australia.
- On passage, they are commonly seen in Japan, Korea and Borneo.[22]
- Even though they are only six to eight weeks old, the baby chicks know how to find their way on their very first migration! They have a special built-in sense of direction and distance.
- An essential stopover on their **migration** south is the Yellow sea, bordered by China, Russia, and the Korean peninsula.

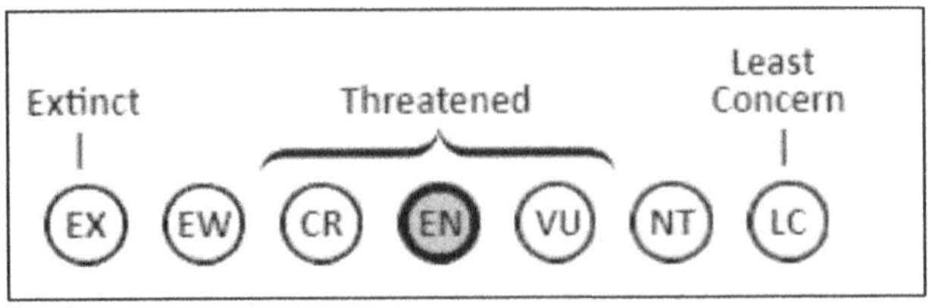

[1]

EPBC List = CE – Critically endangered

Plains wanderer

Pedionomus torquatus

- These shy little birds are found on the bare saltbush plains or in the grasslands and herb fields of northern Victoria and outback New South Wales, stretching into Queensland.[23]
- Plains wanderers are so shy and **secretive** that even scientists rarely see them. Researchers estimate there are fewer than a thousand of these **unique** birds left in the wild. [23]
- The best evidence of their presence is taken from automatic recording devices sensitive enough to pick up their gentle 'ooo' calls at **dusk** and dawn. These recording devices have recently become more silent, indicating that the Plains-wanderer is struggling to survive. [23]
- The Plains-wanderer is an avian oddball, a mysterious little grassland specialist. There are no other birds like them in the world. Elusive and shy, these birds are most often encountered spotlighting at night, startled from their **roost** on the ground. They crane up on tiptoe when alerted to danger, then crouch down to make a dash for cover.
- Standing about 12-15cm tall with a wingspan of 28-36cm and weighing less than 100 grams, Plains-wanderers are small fawn-coloured quail-like ground-dwelling birds. Their dappled feathers include white and blackish marks, with spots and streaks on the head and neck. The larger female is easily distinguished by her **prominent**, white-spotted diamante collar above a rich **rufous** breast patch.
- Plains wanderers live in semi-arid, lowland **native** grasslands. Their nests are shallow, grass-lined scrapes that are scratched into the ground. They **forage** during the day for a wide variety of seeds and ground-dwelling insects and do not require regular access to water bodies; instead, they fulfil their hydration requirements through the food they **consume** and via pecking dew and raindrops from leaves. Each bird is **solitary** and has an average home **range** of around 12 hectares. Males and females with overlapping **ranges** form breeding pairs, with the larger females **defending** their territories and mating with several birds in a season, while the males **incubate** eggs and raise the young.[9] In favourable conditions, females can lay multiple **clutches** of three to five eggs per year.
- **Predation** is a danger that is ever-present from dingoes, foxes, **feral** cats, and falcons - the **species**' ground-nesting habits, poor flying ability, and tendency to run rather than fly from **predators** make it easy **prey**.
- Most **significant** is the loss of **habitat** to farms. Over and **undergrazing** by **introduced** and **native herbivores** and invasion by weeds significantly decrease **habitat** quality. Other **threats** include the impacts of altered **fire regimens**.
- The **survival** of the Plains wanderer will depend upon collaboration between zoos, farmers, scientists, wildlife organisations, and broader climate change mitigation action. [23]

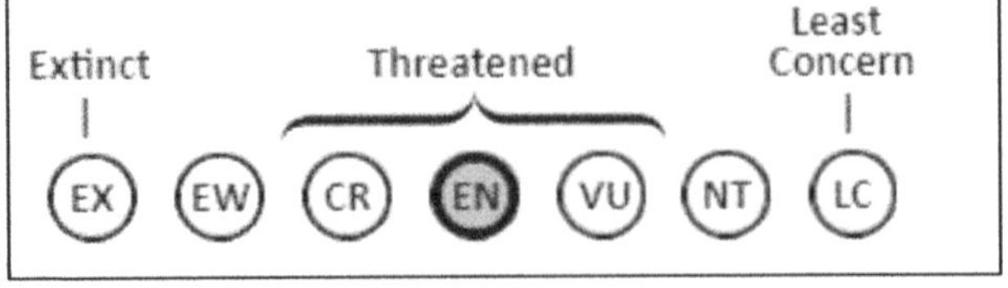

[1]

NCA List = CR = Endangered

EPBC List = CE = Critically Endangered

Swift parrot

Lathamus discolor

NATIVE TO AUSTRALIA

- The swift parrot is a slim, medium-sized parrot. As its name suggests, it has a very fast and direct flight. In flight, its body is **streamlined** with angular and pointed wings; it has a long, pointed purple-red tail. The body is mostly shades of very bright green, with a darker blue patch on the crown of its head. The forehead to the throat is bright crimson red, and there is a crimson patch at the bend of the wing. The male is slightly brighter in colour than the female.
- They are noisy, active and have a striking appearance. When feeding, the swift parrot is very **agile**, a bit of an **acrobat**, often hanging upside down.
- The swift parrot breeds in Tasmania and **migrates** to the mainland for winter, **foraging** on flowering eucalypts mostly in open ironbark forests.
- They feed in the **canopy** of flowering eucalypts, mainly eating **nectar**, but sometimes seeds and flowers, only coming down to drink.
- In dry years when flowering eucalypts do not produce enough flowers for them to feed on, swift parrots will travel greater distances to find food. Some birds may fly as far as coastal south-east Queensland.
- The swift parrot undertakes the longest parrot **migration** in the world, **migrating** between Tasmania and southeastern mainland Australia, covering distances up to 5,000 k!
- When in mainland Australia, they sometimes get together in big **flocks** and can be seen with smaller honeyeaters and lorikeets.
- They are usually **nomadic**, meaning they may spend only a few hours or weeks and sometimes months in some places. They **roost** at night as a group, often in the same trees.

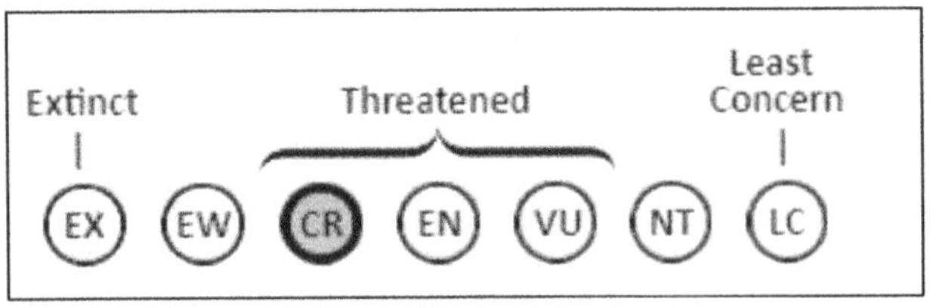

[1]

EPBC List = CE = Critically endangered

Greater bilby

Macrotis lagotis

- The greater bilby is an Australian **marsupial** known for its long ears, digging abilities, and **unique adaptations** to **arid environments**. These **nocturnal** animals don't drink water, relying on the **moisture** from their diet of insects, seeds, fruits, and bulbs, and their backward-facing pouches keep them clean while digging extensive burrows.
- Bilbies are found in dry, desert areas and dry forests of Australia and are well-suited to hot, **arid** conditions.
- Bilbies have a pointed snout, long pinkish eats, a long black tail with a white tip, and strong forelimbs and claws for **burrowing** and digging up food.
- Males are much larger than the females and can be up to 2.5kg in weight.
- Each bilby moves between up to 12 burrows in their home range in response to the availability of food. The burrows may be 3m long and spiral down about 2m.
- They are strictly **nocturnal** so active at night, spending their days sheltering from **predators** and heat in their burrows and emerging at night to find food.
- Bilbies don't need to drink water because they get all the moisture they need from the food they eat.
- Bilbies usually live on their own, but sometimes form pairs or small groups including the young. Female bilbies have pouches that face backwards, preventing dirt from getting in when they dig.
- They have very good senses of smell and hearing, and retreat to their burrows if alarmed. Their large, hairless ears help them have excellent hearing to detect predators and also help them cool down in the heat.
- Bilbies have poor eyesight and rely more on their sense of smell and hearing to find food and sense danger.
- They feed on seeds (which are licked up by their long, slender tongues), fruit and insects (e.g. termites, ants and beetles), insect **larvae** (e.g. witchetty grubs), bulbs and **fungi**. Areas where bilbies are active may be pockmarked with shallow holes, up to 10 centimetres deep, where they have been digging for food.
- In Australia, <u>chocolate bilbies are often used in Easter baskets instead of Easter bunnies</u>.
- The massive amount of soil bilbies turn over while digging helps the environment, making them important "ecosystem engineers".
- Unfortunately, bilbies are a vulnerable species in the wild due to factors like introduced predators like foxes and feral cats.

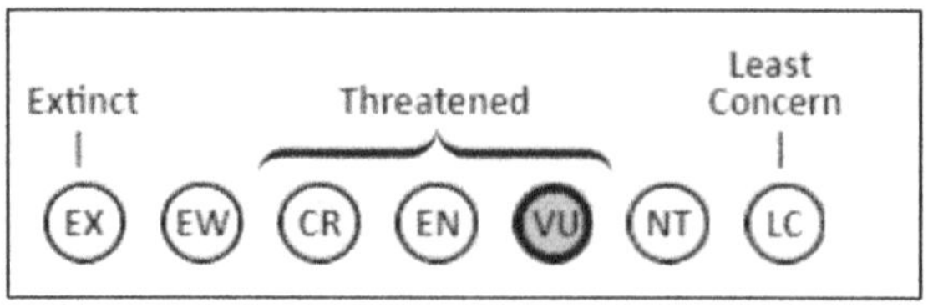

[1]

EPBC List = V = Vulnerable

Brolga

Grus rubicunda

- Brolgas are large, graceful birds found in Australia and New Guinea, known for their impressive size, unique dances, and their special place on the Queensland coat of arms.
- In northern Queensland they can easily be confused with the Sarus cranes which have more extensive red on their heads and necks, and pinkish legs. To further confuse identification, two hybrid species have emerged in northern Australia called the Sarolga or Antigone.
- Brolgas are one of Australia's largest flying birds, standing up to 1.4m tall with a wingspan of over 2m. They have a pale grey body, long legs, and a distinctive red to orange patch on their head with a black "dewlap" (hanging skin) under their chin.
- Brolgas are famous for their elaborate **courtship** dances, which involve leaping, head shaking, and loud trumpeting calls, helping to strengthen their pair bonds. These dances are so iconic that they are featured in Aboriginal legends and dances.
- You can find brolgas across tropical northern Australia. There is a significant population in Victoria and New South Wales, though their conservation status varies by region.
- They inhabit shallow open wetlands, coastal mudflats, and irrigated farmlands. They need these wetland areas for feeding and breeding.
- Brolgas are **omnivores**, meaning they eat both plants and animals. Their diet includes wetland plants like sedge **tubers**, as well as insects, **amphibians**, and even small animals like mice. They are very skilled at **foraging**, even with their heads submerged underwater.
- The brolga is a significant symbol for Queensland, as it was added as a supporting animal to the state's coat of arms in 1977 and officially named the state bird in 1986.
- Brolgas are thought to mate for life, and both parents work together to **incubate** eggs and care for their chicks for up to a year or more.
- Brolgas are unique among cranes as they possess a gland near their eyes that helps them get rid of excess salt, allowing them to live near saltwater marshes and even drink **saline** water. This specialised salt gland does not occur in Sarus cranes, which appear to rely on fresh water sources.
- Brolga populations have declined due to **habitat loss** and **modification**, with specific conservation efforts underway in various parts of the basin.

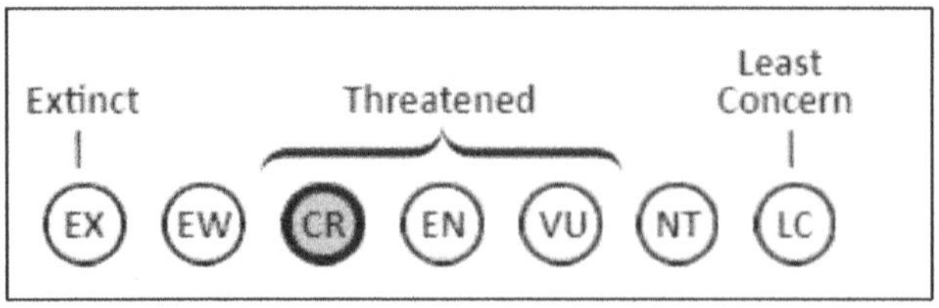

[1]

EPBC = Not listed
Conservation Status:
NSW,South Australia = Vulnerable
Victoria = Endangered

Eastern ground parrot LOCALLY EXTINCT IN THE MURRAY-DARLING BASIN

Pezoporus wallicus

- The Eastern ground parrot is a shy, ground-dwelling parrot with distinctive grass-green **plumage** marked with black and yellow barring, a red band on its forehead, and a long tail, known for its unique whistling call and its preference for coastal heathlands and **sedgelands**. The last sighting recorded in the MDB was in The Coorong in 1941[23]!
- It is one of only a handful of ground-dwelling parrots globally and is **adapted** to a life on the ground, using its long legs and feet for walking and running through **dense vegetation**.
- Adults are around 30cm long, with their most recognisable features being the grass-green feathers with black and yellow barring on the body and wings, and a characteristic red band above the bill.
- Their call is a distinctive high-pitched whistle that ascends in pitch, often heard at dawn and dusk, and is a more reliable way to locate them than visually sighting them.
- They are highly dependent on coastal heathland and **sedgelands** for food, nesting, and shelter.
- Interestingly, they can escape fires and often return to their preferred habitats 8-10 years after a fire when the vegetation reaches peak growth. However, the exact refuge during fires is a mystery.
- They are primarily **granivores**, feeding on seeds from a wide variety of **sedges**, grasses, herbs, and shrubs, and they require **dense vegetation** to **forage** effectively.
- While the global classification is "Least Concern" due to the inclusion of the Tasmanian subspecies, mainland populations, especially in Victoria, are classified as endangered or vulnerable due to **habitat** loss and **fragmentation**, including impacts from the 2019-2020 bushfires.
- Eastern ground parrots are typically **solitary** or found in pairs, though younger birds may form small flocks before dispersing to find their own **territories**.

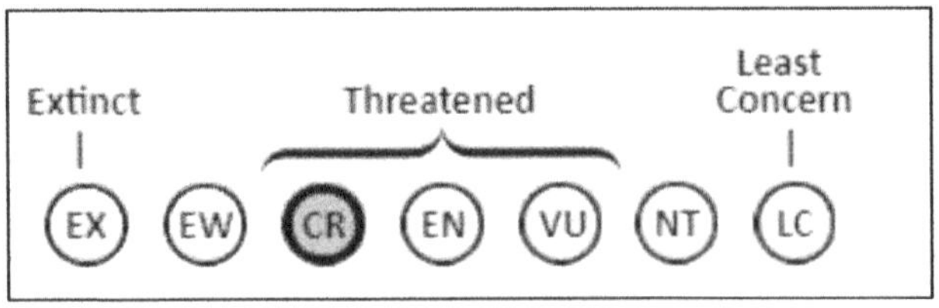

[1]

EPBC = Not listed
Conservation Status:
Victoria = Endangered

Galaxiidae - Galaxias

- Several *Galaxias* **species** occur in the Murray-Darling Basin, including the mountain galaxias, flat-headed galaxias, barred galaxias, obscure galaxias, ornate galaxias, riffle galaxias, spotted galaxias, and stocky galaxias. Galaxids are also known as Jollytails!
- In Australia 22 **species** in 5 **genera** are recognised.
- The mountain galaxias is the most **abundant** and widespread, found in **montane** and upland areas, while the flat-headed galaxias is known only from the southern half of the Basin and prefers still or slow-flowing waters. The barred galaxias has experienced a significant decline and is now restricted to the headwaters of the Goulburn River catchment.
- The spotted galaxias is present in the upper Campaspe and Loddon drainages, likely as a result of translocation, though a historical record suggests a wider distribution might have existed.
- Galaxias fish, found mostly in the Southern Hemisphere, are a **diverse** family of freshwater fish, with some **species** having unique **adaptations** like climbing waterfalls and others facing challenges from introduced **species**.
  Most Galaxias fish prefer the cold water found in streams, rivers, and lakes, especially in **temperate** areas of Australia.
- While some Galaxias stay in freshwater their whole lives, many, like the common Galaxias, have a special journey called **amphidromous migration**. Their babies **hatch** in rivers and then float down to the ocean, only to return to the rivers as young fish to grow up.
- The climbing galaxias is famous for its ability to climb up damp rock faces and even waterfalls, using its special fins and wriggling its body like a lizard.
- Galaxias fish munch on all sorts of things like water insects, tiny shrimp, and even flies and beetles that fall into the water.
- Unlike many other fish, Galaxias fish don't have scales. Their bodies are smooth.
- The common galaxias often migrates downstream to lay its eggs during new or full moons in autumn. The eggs are small and need careful conditions to hatch, and sadly, can be eaten by mice or spiders.
- Many Galaxias are in danger because of introduced fish like trout. Trout eat Galaxias and compete with them for food, causing their numbers to drop.

NOTE: The above information is of a general nature. To find out more about the IUCN and EPBC status of Galaxias fish, visit the IUCN Red List and/or Atlas of Living Australia.

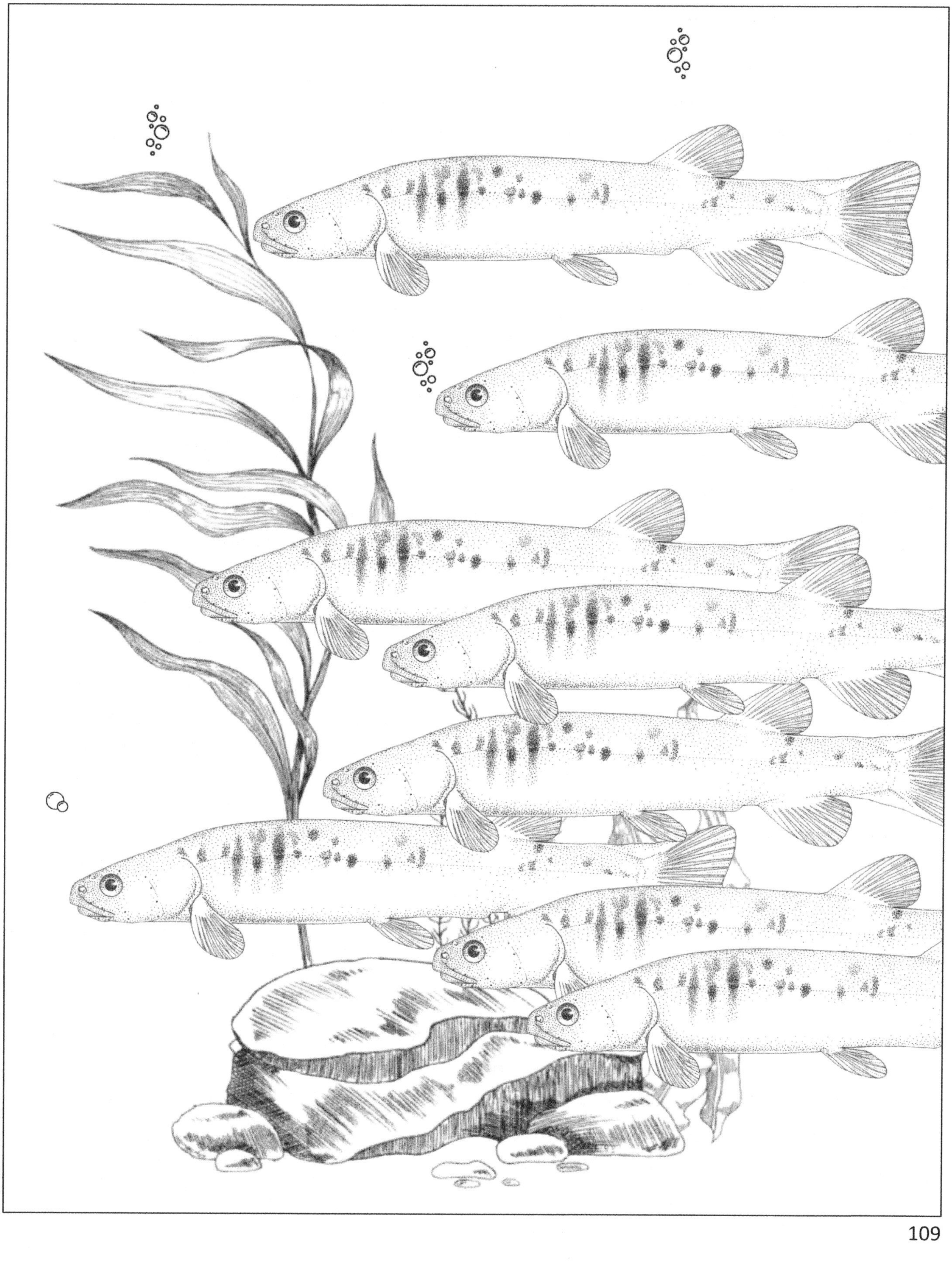

White throated needletail

Hirundapus australis

- White-throated needletails are large swifts with a **robust**, barrel-like body and long wings. The tail is short and square, with the protruding feather shafts giving a spiky, needle-like appearance.
- They are greyish-brown except for a white throat and a white patch extending from the base of the tail to the **flanks**.
- They can fly very fast - they are thought to reach speeds of 170km/h in **horizontal** flight! That is REALLY fast – <u>almost twice as fast as a cheetah</u>, which can reach speeds of 95km/h! The record for the fastest human running speed is 44.72km/h.
- White-throated needletails spend almost all their time in the air. For a time, it was commonly believed that they did not land while in Australia; it has now been observed that birds **roost** in **tree canopies** with **dense** leaves or in hollows, and **radio-tracking** has since confirmed that this is a regular activity.
- White-throated needletails grow up to 20cm long and weigh up to 115-120g. They have a **lifespan** of about 8.5 years
- White-throated needletails breed in Central Asia and southern Siberia and **migrate** to winter south in the Indian Subcontinent, Southeast Asia, and Australia. They are **rare vagrants** in Western Europe and have been recorded as far west as Norway, Sweden, and Great Britain.
- In Australia, they are found in coastal areas of Queensland, NSW, and the western slopes of the Great Dividing Range; they also travel to Victoria, Tasmania and south-eastern South Australia.
- They **forage** over forests and grasslands in both mountainous and lowland areas, feeding on flying insects such as termites, ants, beetles, and flies. They catch the insects in flight in their wide beaks, usually feeding in rising **thermal currents** associated with **storm fronts** and bushfires. They are commonly seen moving with wind fronts; hence, they are also known as the storm bird! While feeding, they protect their eyes with a special **membrane** and a small ridge of feathers.
- They also drink while flying by **skimming** the surface of the water....they are able to do this because they have wide, gaping beaks.
- White-throated needletails communicate with each other using loud, sharp calls, chirps, a piping sound, and a churring twitter. They are also **social** birds and often form **flocks**.
- **Threats** to the White-throated Needletail include:
    - crashing into wind turbines, overhead wires, windows, and lighthouses
    - **insecticides** that reduce their food supply or cause secondary **poisoning**
    - loss of **habitat**, which means fewer places to **roost** and possibly less food
    - logging and hunting in the areas where it breeds in the Northern Hemisphere.[30]

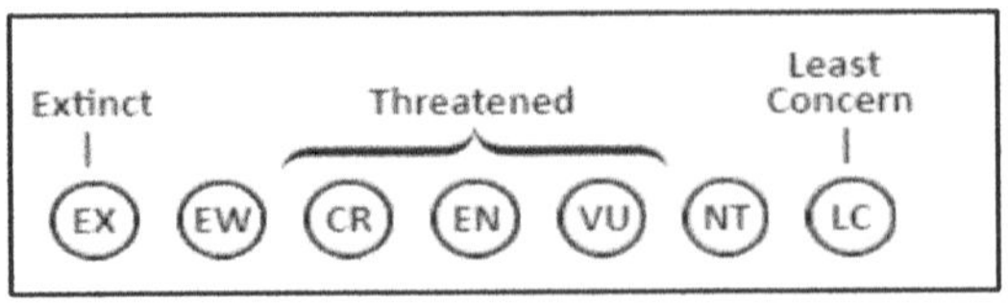

[1]

EPBC List = V = Vulnerable

Ruddy turnstone

Arenaria interpres

- These birds are found along coastlines worldwide and migrate to breeding grounds in the Northern Hemisphere, then travel south for the non-breeding season, including to Australia from September to May.
  Ruddy Turnstones get their name because they use their bills to flip over stones and seaweed to find yummy insects, **crustaceans**, and other small creatures hiding underneath.
- They are global travellers and migrate very far! Some Ruddy Turnstones fly from Australia to Taiwan in just about six days, covering thousands of miles. They are capable of impressive non-stop journeys of up to approximately 7600km.
- They are amazing divers and strong fliers and can even dig holes in the sand to catch **prey** that's buried deeper.
- Their diet is varied, including insects, spiders, **crustaceans, molluscs**, and sometimes even eggs or **carrion**. They will eat maggots in rotting seaweed and sandhopper **larvae**.
- Ruddy Turnstones have different looks for different times of the year. They are chunky birds with short orange legs and bills, and distinctive plumage that changes between breeding and non-breeding seasons. In breeding plumage, males have a bright reddish back, black chest patches, and a black and white face, while females and winter plumage are duller.
- Their calls include a staccato "tuk-a-tuk" or a warbling "tuk-tuk-tuk-tuk".
- Ruddy Turnstones are **monogamous** and usually return to the same mates and breeding grounds each year. Both parents care for the chicks.though the female departs before **fledging**.
- Ruddy Turnstone chicks are tough - they can leave the nest soon after **hatching** and are able to feed themselves, though they can fly after about a month.
- Outside of breeding season, they often gather in large groups, sometimes mixing with other types of shorebirds, but they can be quite **aggressive** when competing for food or defending their **territory** during breeding season.

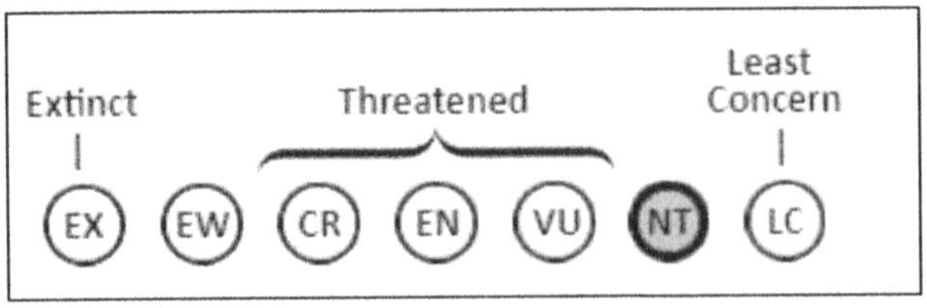

[1]

EPBC = V = Vulnerable

Small purple pea

Swainsona recta

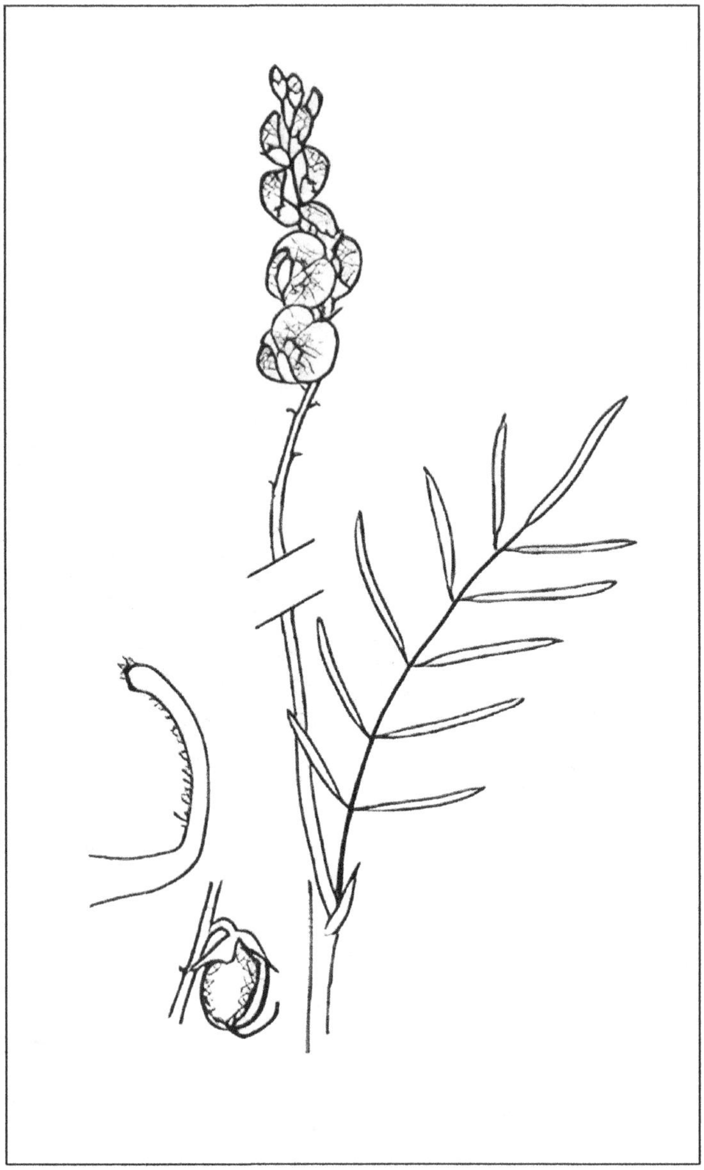

- The small purple pea is a slender, erect **perennial** herb native to Australia. This plant is characterised by its rigid stems, deep taproot, and clusters of bright purple, pea-shaped flowers with distinctive white spots on the standard petal.
- It is a **perennial** plant that overwinters as a woody rootstock and re-sprouts between April and August, with flowering occurring in spring.
- The plant possesses a substantial taproot that can extend up to 60cm deep into the soil, likely aiding in water and **nutrient** acquisition.
- The likelihood of flowering is influenced by the number of frost nights in the previous season, with more frost (7-15 nights at or below -4°C) promoting flowering.
- Fire can also play a role in its life cycle by stimulating re-sprouting and seed **germination**, although populations have declined due to **habitat** loss and low recruitment rates.
- The seeds are small, hard-coated, and kidney-shaped, and the plant is a **legume**, meaning its roots have nodules containing soil **bacteria** that fix atmospheric nitrogen, benefiting its **nutrient** requirements.
- Swainsona recta is listed as endangered in NSW and Victoria, and **critically endangered** under the Victorian Flora and Fauna Guarantee Act 1988.
- It is a key species within the **critically endangered** Box Gum Grassy Woodlands **ecological** community, often found in association with trees like Blakely's Red Gum, Yellow Box, and Long-leaf Box.
- Significant **conservation** efforts are underway, including seed collection, propagation, and **habitat** management, to ensure the survival of this **unique species**.
-
-

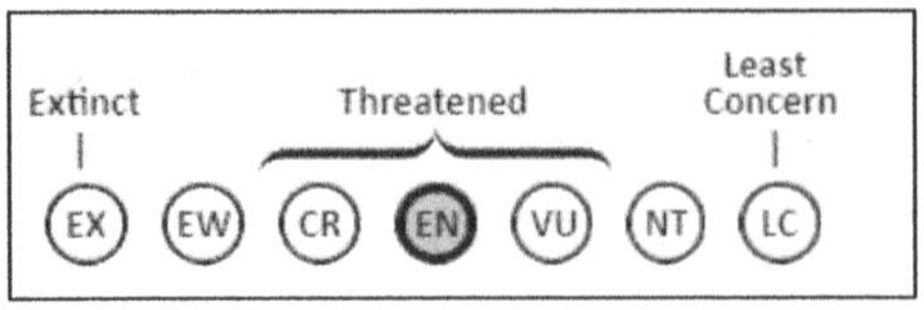

[1]

EPBC = E = Endangered

Duramana fingers

Caladenia attenuata

- **A critically endangered** orchid found in a very limited area of New South Wales, Australia, north of Bathurst. It is considered **critically endangered** at both the state and Commonwealth levels in Australia.
- This orchid has only been recorded in two **localities** within the Bathurst-Ilford region of NSW, with an estimated area of occupancy of only 8km^2.
- Unlike many orchids, Duramana Fingers grows on an unusually tall spike, reaching up to 24cm high.
- It features a single hairy leaf and has white flowers, which are typically less than 10mm across, with a **dorsal sepal** that loosely hoods the column and a wrinkled midlobe of the **labellum**.

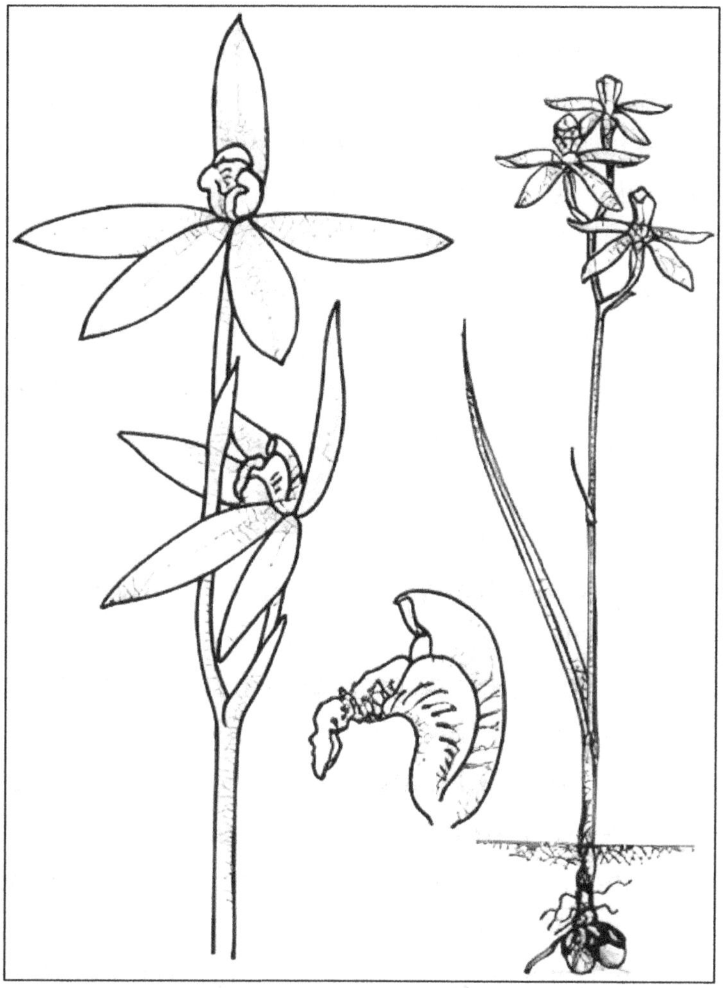

- The Duramana Fingers typically flowers in the spring, usually between October and November.
- Major threats to its survival, like many native species, include grazing by cattle and rabbits, and competition from **invasive weeds**.
- Discovered and formally recorded only in the 1960s, very little is known about this **species** due to its extreme **rarity**.
- The Duramana Fingers is considered "data-deficient" by programs like Saving our Species (SoS), meaning a specific **conservation** project cannot yet be developed for it.

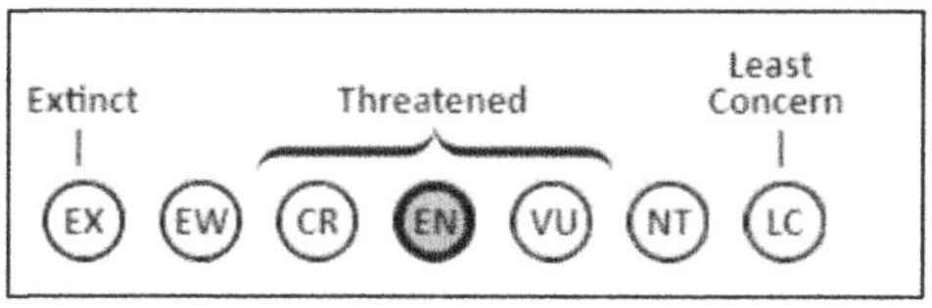

[1]

EPBC = CE = Critically endangered

Metallic sun orchid

Thelymitra epipactoides

- The metallic sun-orchid, is a nationally **endangered** Australian orchid known for its metallic appearance, variable flower colours, and sensitivity to habitat fragmentation. This large sun-orchid, typically reaching 21-52cm in height, produces a single, fleshy, tubular leaf and 7-31 flowers with a metallic sheen due to the proportion of reflective cells in its petals and **sepals**.
- Its name comes from the **iridescent**, metallic sheen on its flowers, which can range in colour from brown to copper, blue, and green.
- It's the largest of the sun-orchid family, growing from a subterranean **tuber** and emerging annually with a single, upright leaf that can be up to 51cm long in mature plants.
- Flowers typically bloom from August to November and open only on warm, sunny days with low **humidity**, releasing seeds in December before the plant becomes dormant for the summer.
- Unlike many other orchids, *Thelymitra* species have similar **sepals** and petals, with the column being the primary feature for identification.
- Native bees **pollinate** it, and the flowers produce a sweet scent when they open.
- The genus name *Thelymitra* is derived from the Greek words for "woman" and "headdress" or "turban," referring to the hooded structure of the column, a unique feature of orchids.
- It is listed as nationally **endangered** and is found in southern Victoria and southeastern South Australia, with **critical** populations in specific regions of South Australia, such as the Murraylands and Riverland.
- Major threats include **habitat fragmentation**, grazing, **weed invasion**, and fire.

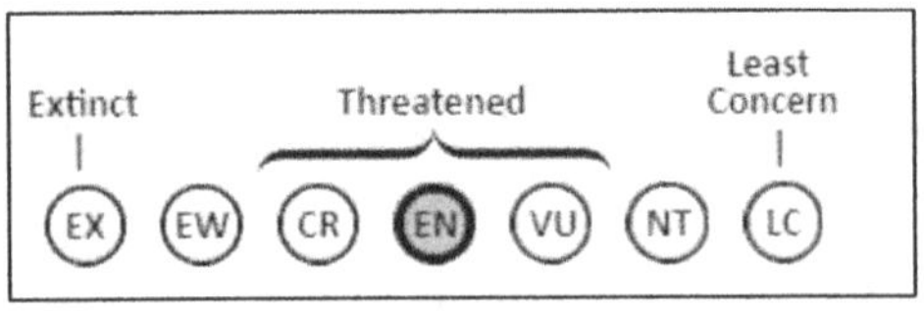

[1]

EPBC = E = Endangered

Rusty Menzies, Rusty Kennedia

Kennedia retrorsa

- *Kennedia retrorsa* is a **critically endangered** and **vulnerable** Australian native climbing herb prized for its vigorous growth and attractive pink-purple, pea-shaped flowers, with a restricted distribution primarily in the Goulburn River Valley of New South Wales.
- It is listed as vulnerable under the EPBC Act and in New South Wales, and critically endangered globally due to its restricted distribution and sensitivity to fire.
- *Kennedia retrorsa* is primarily found in the Mt Dangar and Goulburn River Valley areas near Putty, NSW, occurring within Goulburn River and Wollemi National Parks and on private land.
- This vigorous climber or mat-like herb is covered with rusty or white hairs, featuring trifoliate leaves and distinctive bright pink-purple or scarlet pea-shaped flowers with a white spot on the upper petals.
- *Kennedia retrorsa* is considered fire sensitive, and fire intervals of less than 10-15 years can lead to subpopulation declines. Its **life cycle** depends on the production and **germination** of viable seed, and it is not known to reproduce vegetatively.
- The species produces seeds in **densely** hairy, flattened pods, and the hard-coated seeds are prone to dormancy, accumulating in the soil.
- It's considered a successional species in some **localities**, meaning it may be **abundant** after fire and then decline due to competition and shading from trees and shrubs.
- Threats include **habitat modification, weed invasion**, reduced water quality, and increased **sedimentation**.
- **Conservation** efforts focus on surveying potential **habitat**, collecting seed, researching seed viability and fire response, and developing **habitat** management strategies, including fire and **weed** management.

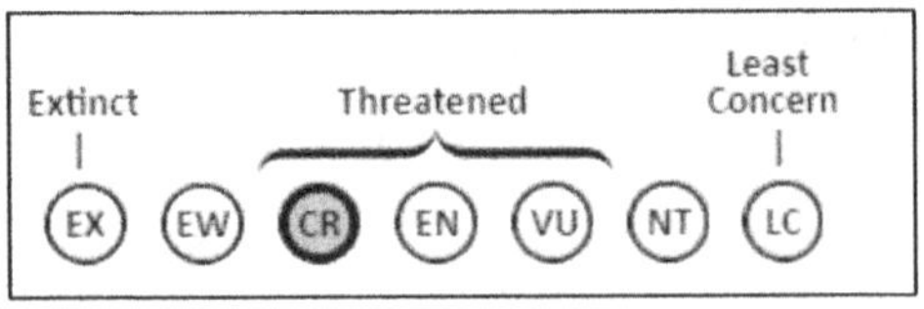

[1]

EPBC = V = Vulnerable

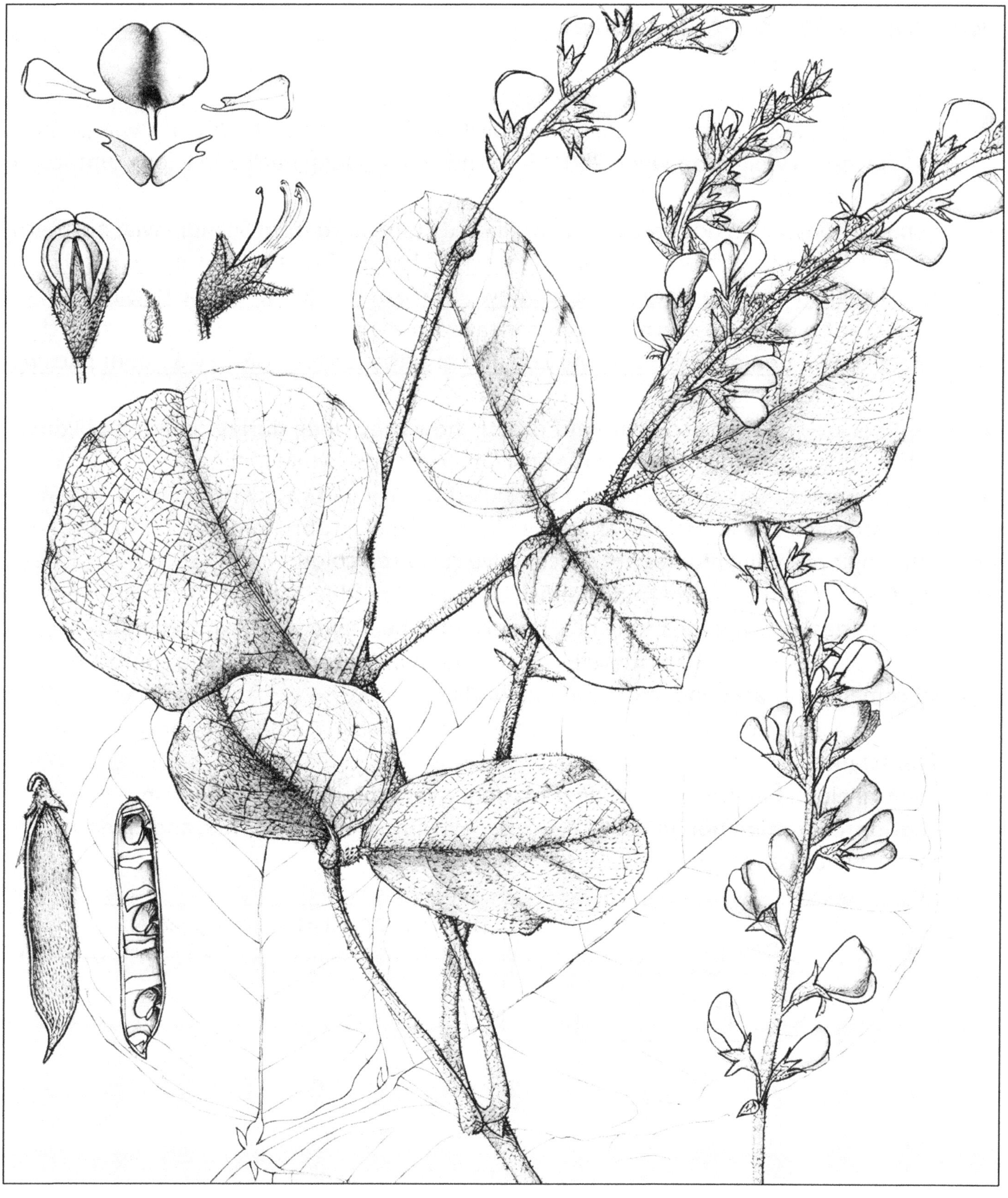

Western Alaskan bar-tailed godwit

Limosa lapponica baueri

- Bar-tailed godwits are large wading shorebirds with a length of 37-39cm, a wingspan of 62.75cm, and a weight between 250-450g. They have a long neck and a long, upturned bill with a dark tip.
- Bar-tailed godwits do not breed in Australia but come here from Scandinavia, northern Asia, and Alaska for the winter. Satellite tracking has confirmed that some birds fly over 11,000km and one tracked godwit flew 29280km in a round trip lasting 174 days!
- The **migration** of the western Alaskan bar-tailed godwit across the Pacific ocean is the longest known non-stop flight of any bird and also the longest journey without pausing to feed by any animal.
- Scientists are still trying to figure out exactly how they sleep during their long flights, but they think they might take very short "microsleeps" while gliding.
- To fly for so long without eating, the Bar-tailed Godwit can shrink its digestive organs, like its stomach and intestines, to save weight and make room for extra fat for the long flight. Their breeding **plumage** is a distinctive rusty red colour.
- They feed on bristle-worms and **shellfish** on coastal mudflats, beaches, mangroves, and estuaries, probing their long bills rapidly into the shallows or exposed mud.
- Feeding parties can be 30 or more birds, including **migratory** adults and some birds that remain all year round, mostly young birds.
- If you see a group of Bar-tailed Godwits together, you can call them a "prayer" or an "omniscience".
- Threats include **habitat** loss and **degradation**, **overexploitation** of **shellfish**, **pollution/contamination** impacts, diseases, extreme weather events, and climate change, which can raise sea levels.

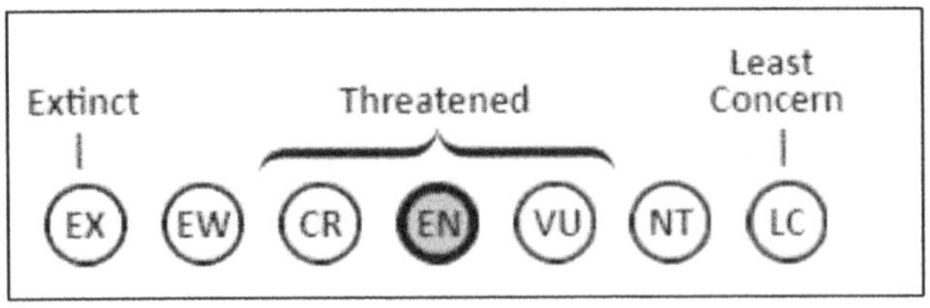

[1]

EPBC list = E = Endangered

Grey-headed flying fox

Pteropus poliocephalus

NATIVE TO AUSTRALIA

- The grey-headed flying fox is amongst the largest bats in the world. They are commonly known as fruit bats or flying foxes and are also the largest bats in Australia.
- They eat fruits and other plant matter, occasionally also eating insects, which they can find due to their keen sense of smell.
- They are mostly **nocturnal**.
- Six flying fox species have already been made **extinct** (due to overhunting)!
- They are important **pollinators**.
- They can also carry diseases such as Australian bat Lyssavirus, rabies, Hendra virus, and Nipah virus.
- They have **cultural significance** to **indigenous** people, appearing in traditional art, folklore, and weaponry. Their fur and teeth were used as a form of **currency** in the past, and some **cultures** still use them as **currency**.
- **Variations** in body weight range from 120 – 1600g. Males are usually bigger than females.
- In terms of threats, flying foxes' **predators** include sea eagles and snakes.
- They also suffer from habitat loss for **foraging**, competition, and extreme weather events. Recent studies indicate that since 1994, over 24,500 grey-headed flying foxes have died due to extreme heat events.

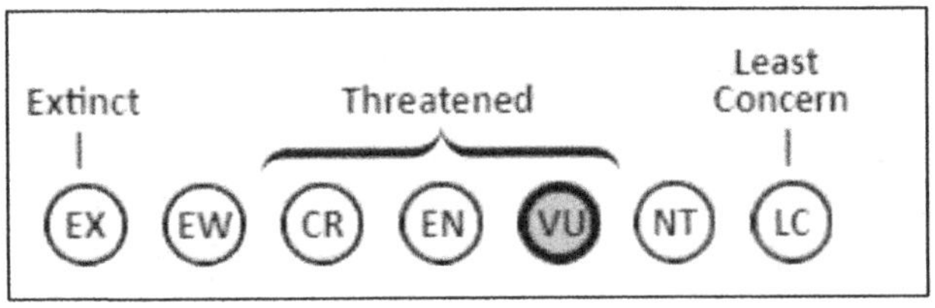

[1]

EPBC List = V = Vulnerable

Hooded Robin

Melanodryas cucullata

- The hooded robin is a medium-sized Australian bird known for its striking black and white **plumage** in males, and its shy, ground-**foraging** behaviour where it pounces on insects.
- They are a type of **passerine** bird.
- They are not normally found in **urban** or **densely** forested areas. Hooded Robins are more positively associated with edge **habitat**s, with deep and sheltered edges
- They are.about 16cm long, <u>about the size of a pen</u>!
- The male has a very distinct black "hood" on its head and neck, with a white chest and belly, and black wings with white stripes. The female is a duller grey-brown, but also has white stripes on her wings.
- Hooded robins are usually seen in pairs or small groups and are quite shy. They are not usually noisy, but males sing in the early morning.
- They are expert hunters! Hooded robins sit on low branches or stumps and then pounce down to catch insects and other small creatures like lizards on the ground.
- The female builds a neat, cup-shaped nest with grass and bark, and uses spider webs to help hold it together. Nests are usually hidden in a hollow or **crevice** in a tree or stump.
- They breed in pairs, typically **monogamous** pairs, during their July to November breeding season. The female **incubates** the eggs, and the male helps out by bringing food and even doing some brooding himself.
- These birds are facing some challenges, mainly due to **habitat** loss from clearing woodlands. Foxes and other **predators** can also threaten them because they **forage** at ground level, becoming easy prey. Nests are within 5m of ground level, which can be within reach of cats.

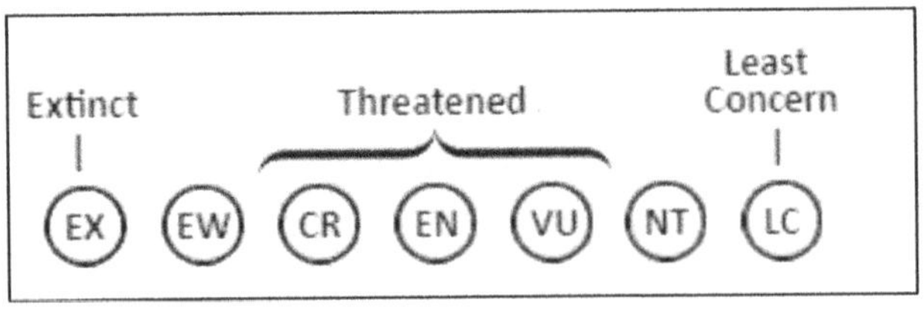

[1]

EPBC List = E = Endangered

Conservation status:
ACT, VIC = Vulnerable

Painted honeyeater

Grantiella picta

- The painted honeyeater is a small, striking Australian bird known for its black, white, and yellow feathers and a pink beak, and for its strong connection to mistletoe plants, which provide both food and nesting sites.
- These **nomadic** birds are often found in eucalypt forests and woodlands with plenty of mistletoe, and their movements and breeding are tied to the availability of mistletoe berries and **nectar**.
- Painted honeyeaters love mistletoe! They eat mistletoe berries when they are ripe, and also drink **nectar** from mistletoe flowers when berries are scarce. They even help spread mistletoe seeds by eating the fruit.
- They build unique, cup-shaped nests that hang from branches, made from things like rootlets, plant fibres, and spiderwebs.
- Painted Honeyeaters are **nomadic**, with their movements dictated mainly by the availability of mistletoe fruits, moving north in winter and returning south to breed in spring and summer when mistletoe fruits are most **abundant**.
- Both male and female Painted honeyeaters help build the nest, **incubate** the eggs, and take care of the baby birds.
- You might hear their call, which sounds like "Georgie", which is a loud, whistled sound that can be reversed and can be heard over long distances.
- While often found in eucalypt and acacia woodlands, they can also be found in urban areas with mature native trees and mistletoe.
- They are considered a **threatened species** in some areas, with threats including **habitat** clearance and extreme heatwaves.

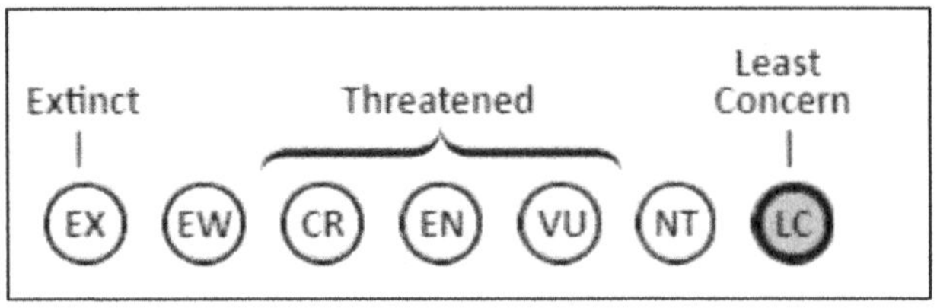

[1]

EPBC List = V = Vulnerable

Grey falcon

Falco hypoleucos

- The grey falcon is a **rare** and **elusive** medium-sized **raptor** found in **arid and semi-arid** Australia, known for its fast aerial hunts and distinctive pale grey and white **plumage**.
- Grey falcons are a compact falcon, mostly grey on top with white undersides and have a yellow **cere** (the fleshy part above the beak) and legs. When they fly, the tips of their wings are darker.
- Adult grey falcons are about 30-45cm long with a wingspan of 85-95cm. It is notable for its reversed **sexual dimorphism**, with females being significantly larger than males (30%), and for being a specialist **predator** of birds, often hunting in aerial pursuits.
- They prefer to live in **arid and semi-arid** environments, open areas with sparse trees in inland Australia, usually where there is less than 500mm of rain each year and near watercourses.
- While they don't **migrate** in the traditional sense, their movements are **nomadic**, driven by the need to find resources and avoid deteriorating conditions.
- They are known for their **agile** and fast flight.
- They are amazing at catching prey by chasing them through the air very quickly. The grey falcon primarily **preys** on other birds, especially parrots and pigeons, and hunts them during fast, level aerial pursuits close to the ground. They also consume small **mammals, reptiles**, **and invertebrates** like locusts.
- You can often hear their calls, which are a clucking or chattering sound.
- The grey falcon is one of the world's **rarest** falcons, and the **rarest** of the six falcons found in Australia, so seeing one is very special.
- Grey falcons don't build their own nests! They use nests made by other birds, like large trees near water.
- You might sometimes see a grey falcon family, which includes an adult pair and a young bird.
- Like many animals, they face threats from things like **feral** cats and changes in their environment, like increased temperatures due to climate change.

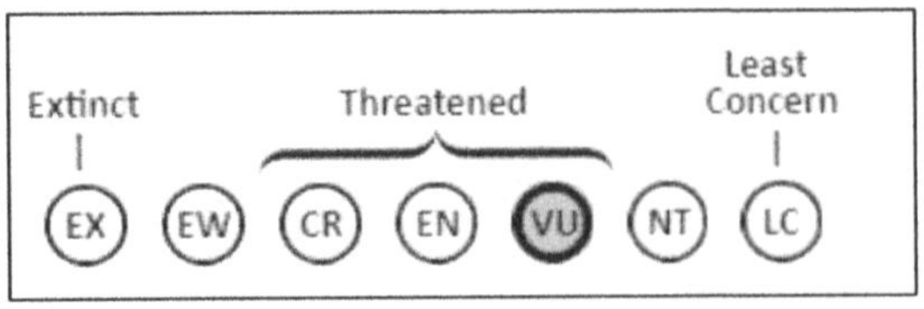

[1]

EPBC List = V = Vulnerable

Southern corroboree frog

Pseudophryne corroboree

- The southern corroboree frog is a strikingly black and yellow-striped **amphibian** found in a small area of the Australian Alps, known for **its unique life cycle**, **toxic** skin, and status as **critically endangered**.
- Southern corroboree frogs are very small, only growing up to 3cm long, about the size of your thumb.
- They have bright yellow or green stripes alternating with black stripes on their top, sides, and legs, with a marbled black, white, and yellow belly, making them very easy to spot.
- They are known to walk rather than jump.
- They primarily eat small ants and other tiny bugs, like beetles and mites.
- Their skin produces a **toxin** that can be **poisonous** to things that try to eat them, which is why they don't have many natural **predators**.
- They breed in **ephemeral** pools and seepages within sphagnum bogs and other wet **habitats**, with males calling during January-February to attract females. Tadpoles overwinter in pools and bogs, staying safe under metres of snow that acts like an insulating blanket.
- They lay eggs in terrestrial nests in mossy areas, and the tadpoles **hatch** when the rain rises to flood the nest in autumn or winter, developing through summer and emerging as froglets in early summer.
- They become inactive during winter, sheltering under logs, rocks, or **dense** leaf litter in the surrounding woodlands and **heathlands**.
- These special frogs only live in a very small area of snow gum woodlands and sphagnum bogs in Kosciuszko National Park in New South Wales.
- Both the Southern and Northern Corroboree Frog species are listed as **critically endangered**, one of the most **critically endangered** frogs in the world, meaning they are at a very high risk of disappearing forever.
- They are threatened by things like a deadly **fungus** called **chytrid fungus**, as well as **habitat** loss and climate change.
- Scientists and zoos are working hard to save them by breeding them in captivity and releasing them back into protected areas.

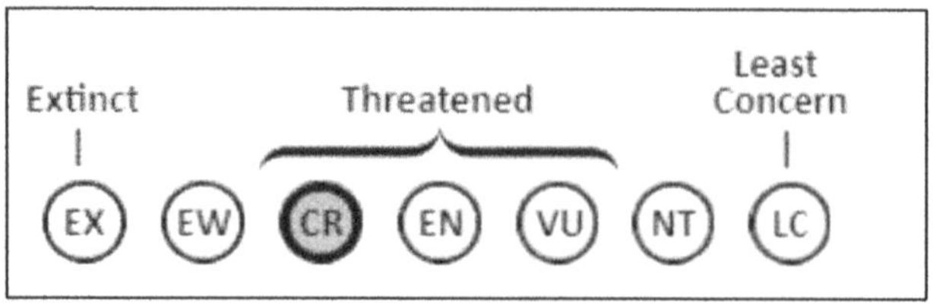

[1]

EPBC List = CE = Critically endangered

Lesser sand plover

Charadrius mongolus

- The lesser sand plover, also called the Mongolian plover, is a small bird found on coastal areas like beaches and mudflats in Australia, known for its impressive long-distance **migrations** and a **unique** feeding method to catch insects, **crustaceans**, and worms. They are highly **social** birds, often seen in large **flocks**, especially when **roosting** during high tide, and sometimes mingle with other shorebird **specie**s.
- Lesser sand plovers are amazing long-distance travellers, **migrating** to Australia for the summer and then flying back to other parts of the world for their breeding season, primarily to Siberia.
- They don't just peck for food; they have a special "run-stop-peck" dance where they run, stop, and then quickly lunge to catch yummy insects, worms, and little crabs hiding in the sand or mud.
- These birds love to hang out with other plovers and shorebirds, forming big groups of over 100 individuals, particularly when **roosting** on sandy beaches, spits and rocky shores.
- You'll most likely see them at the beach, especially on the **intertidal** sandflats and mudflats, which are areas of the beach that get covered by water at high tide and are exposed at low tide.
- Like other plovers, they might use a "false brooding" display to trick **predators** away from their nests if they get too close. They might pretend to be hurt or have a broken wing to lure the danger away from their precious eggs or chicks.
- When lesser sand plover babies **hatch**, they are born with their feathers and can already walk and follow their parents! They can even join their first **migration** within a few weeks.
- The flight call of the lesser sand plover is described as a hard trill.
- Like other shorebirds, they are sensitive to changes in their **habitat**, so it's important to keep beaches clean and protect their nesting areas.

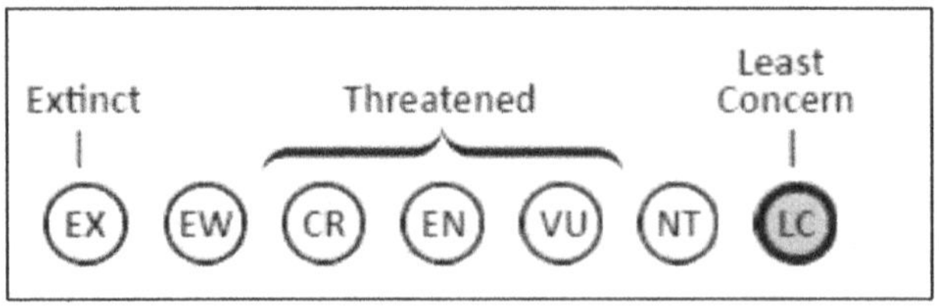

[1]

EPBC List = E = Endangered

Platypus

Ornithorhynchus anatinus

- Platypuses are **unique, semi-aquatic mammals** found in Australia and Tasmania. They are known for their duck-like bill, beaver-like tail, and the fact that they are one of the few mammals that lay eggs. Male platypuses also have **venomous** spurs on their hind legs.
- Unlike most mammals that give birth to live young, platypuses are **monotremes** and lay eggs.
- They have a **bill**, not a beak: The platypus's **bill** is made of soft, leathery skin and is sensitive to electrical signals. They can detect the electric fields generated by their prey, helping them to locate food in murky water.
- They feed on insect larvae, freshwater shrimps, worms and yabbies, which they bring to the surface to eat.
- Male platypuses have a venomous spur on their hind legs, which they use to defend themselves and during mating season.
- Platypuses have two layers of fur for insulation and waterproofing. Platypuses use their fur to trap a layer of air next to their skin so they can remain buoyant and dry when they're underwater, which they are a lot. The platypus spends about 12 hours every day underwater looking for food.
- They don't have teeth; instead, they grind their food with hard plates in their **bill**. They have no stomach, and their **oesophagus** connects directly to their intestines.
- They are **semi-aquatic** and live in and around rivers and lakes, spending a lot of time swimming and **foraging** for food.
- Baby platypuses are called **puggles**. Platypus **puggles** don't stay **puggles** for long, though. They typically leave their nesting **burrows** after four months, after which they are referred to as **juvenile** platypus.
- They are most active at night, searching for insects, worms, and **crustaceans**.
- Platypuses are excellent swimmers. They use their webbed feet and powerful limbs to **navigate** underwater.
- They have 10 sex chromosomes. Unlike humans, who have two sex chromosomes (XX or XY), platypuses have 10, five X and five Y.
- They can dive for extended periods and can hold their breath for up to 10 minutes and dive to depths of up to 1 metre.
- They have biofluorescence; under UV light, platypuses can glow green and blue, a trait shared with other **mammals** like wombats and some flying squirrels.
- Platypuses are found only in eastern Australia, **inhabiting** rivers and streams.
- Platypuses are currently listed as "near threatened" due to **habitat loss** and other factors.

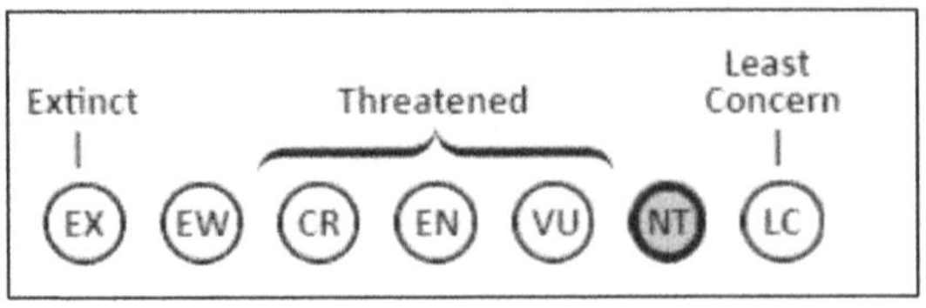

[1]

EPBC List = Not currently listed

Conservation status:
VIC = Vulnerable SA = Endangered

Red-necked avocet

Recurvirostra novaehollandiae

- The red-necked avocet, endemic to Australia, is an elegant shorebird known for its distinctive chestnut-brown head and neck, white body, and long, upturned black **bill**. It is a **nomadic** species that **inhabits** freshwater and saltwater wetlands and **estuaries** across mainland Australia, moving in small **flocks** in search of suitable **habitat**, especially in response to rainfall.
- Red-necked avocets are also known as the Australian avocet, cobbler, cobbler's awl and painted lady.
- Red-necked avocets feed by wading in shallow water and sweeping their upturned **bills** side-to-side just below the surface to find **aquatic** insects, **larvae, crustaceans**, and seeds. They will also sometimes swim and up-end to feed.
- These birds are highly **nomadic**, readily moving in small **flocks** to find suitable feeding grounds, especially after rainfall, and the availability of water in inland areas.
- During **courtship**, red-necked avocets engage in bowing dances and **bill**-crossing in shallow water, with males typically having a more sharply upturned **bill** than females.
- They breed in loose **colonies**, typically from August to September, although this can vary. Their nests are shallow scrapes lined with water plants in which they lay a single **clutch** of three to four eggs.
- Their call is often described as a "toot-toot" sound, similar to a clown horn or toy trumpet, and they also have sharper yelping notes for warning calls.
- Red-necked Avocets are generally **social** birds, found in loose **colonies** during breeding and small **flocks** outside of the breeding season, and they are known to associate with black-winged stilts.
- While relatively large, their eggs and chicks can be **vulnerable** to introduced **predators** like red foxes and feral cats.

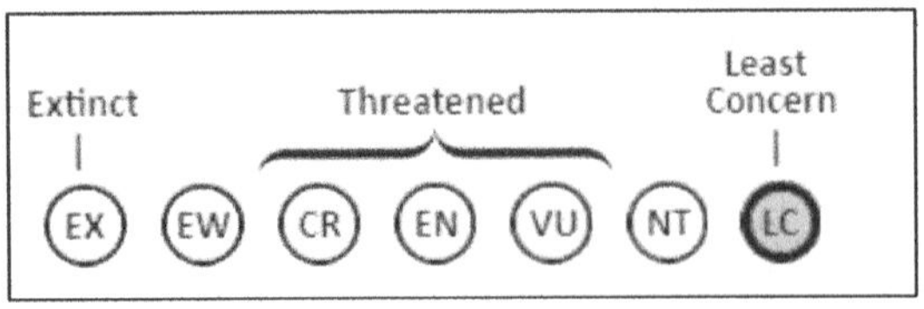

[1]

EPBC List = Not currently listed

Dusky antechinus

Antechinus swainsonii

- Dusky antechinuses are tiny, mouse-like **marsupials** that are very active, eating insects, spiders, and sometimes lizards. A fascinating fact is that after a short mating period, almost all male dusky antechinuses die, overwhelmed by the stress of mating, while the females live to raise their young.
- Dusky antechinuses are the largest species of antechinus and can be from 12-31cm long. When fully grown, they weigh between 16 and 170g.
- There are three **subspecies** of dusky antechinus recognised, and the dusky antechinus can be distinguished from its relatives by its much darker fur, which is also apparent in the pale form.
- The largest and darkest fur dusky antechinus occur in higher **altitudes**.
- They are small, greyish-brown, mouse-like animals with pointed heads and large eyes.
- Unlike mice, dusky antechinuses are **carnivores** and mostly eat insects, spiders, and cockroaches. They will also eat small lizards and frogs if they can catch them.
- They are a ground-dwelling/**semi-arboreal mammal** that **forages** for insects, **nectar**, and small **prey**. These animals are active during the day and night, spending most of their time looking for food.
- The males have a very short life. They all die after the mating season in winter due to the stress from mating, a phenomenon called "big bang reproduction"**.**
- Female dusky antechinuses can have **litters** of up to 12 young, which they raise.
- They are known for being unusually vocal, sometimes chattering and hissing.
- They are known to prefer moist, leafy areas, with logs, branches and leaf litter, such as forests, rainforests, and **dense** undergrowth on the ground, where they build nests from leaves.
- Like other animals, they sometimes take short, deep naps called **torpor**, to save energy, which slows down their bodies and reduces the need for food and water.

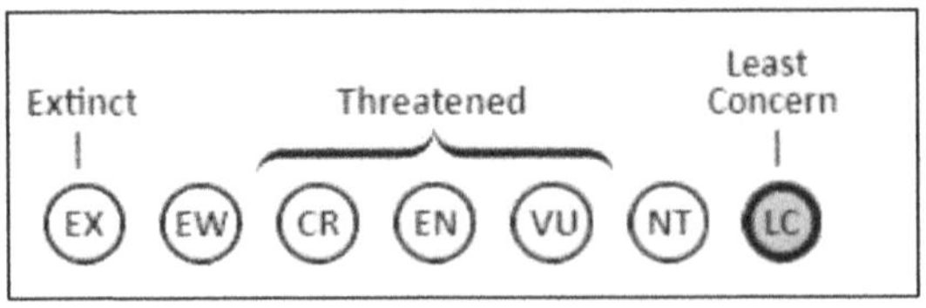

[1]

EPBC List = Not currently listed

Tawny frogmouth

Podargus strigoides

- Tawny frogmouths are fascinating Australian birds known for their excellent **camouflage**, which helps them blend in with tree branches during the day. Despite looking like owls, they are actually related to nightjars.
- They are **nocturnal** hunters, using their wide beaks to catch insects and small **prey**, and both parents share the responsibility of raising their young in their loose, stick nests.
- Tawny frogmouths might look like owls with their big yellow eyes, but they are not owls at all! They are more closely related to nightjars, which are also **nocturnal** birds.
- These birds are amazing at hiding. During the day, they sit perfectly still on a tree branch, making themselves look exactly like a broken branch or a dead log with their eyes closed and beak pointing upwards.
- Unlike owls that have sharp talons for catching **prey**, Tawny frogmouths have wide, strong beaks that they use to catch bugs, mice, and other small animals; they sit and wait for prey to come within striking distance before catching it with their beaks.
- Tawny frogmouths are excellent parents! The mother lays two or three eggs, and both the male and female take turns **incubating** them and feeding the chicks after they **hatch**.
- Even though they are good parents, they are not great at building nests. They just pile up sticks and leaves, and sometimes the chicks or eggs can even fall out.
- They make a low booming "oom-oom-oom-oom" noise, but if they feel threatened, they can hiss loudly at the same time as opening their beaks and eyes wide, appearing larger and more **intimidating**.
- Tawny frogmouths can survive in many different temperatures, even extreme heat by panting like a dog, and cold by going into a short-term **hibernation** called **torpor**.
- They can be found in **urban** areas, often **roosting** on trees in backyards.
- They have special asymmetrical wing feathers for silent flight and longer tail feathers to help them steer while flying.
- Tawny frogmouths stay with the same partner for their entire lives unless one of them dies.

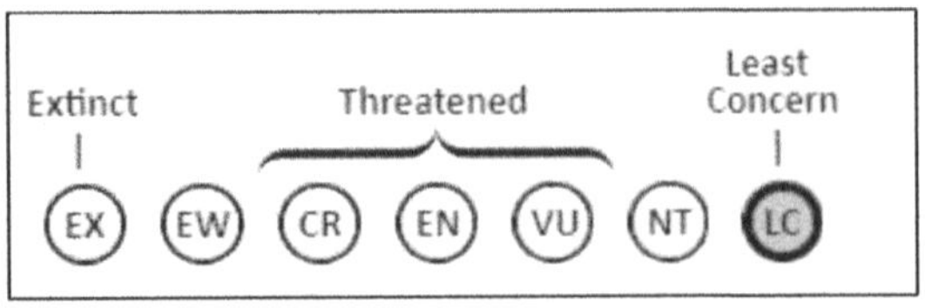

[1]

EPBC List = Not currently listed

Numbat

Myrmecobius fasciatus

- The Numbat was absent in the Murray Darling Basin for over 100 years until 1993, when 15 individuals were released inside a 1,100 hectare **feral predator**-free fenced area at Yookamurra Wildlife Sanctuary on the land of Ngarrindjeri Nation in the south east of Australia. In 1999, another Numbat population was established at Scotia Wildlife Sanctuary on Barkandji country in far western NSW. The reintroduction consisted of 43 founders released inside a 4,000-hectare feral predator-free fenced area. The fenced area was then expanded to 8,000 hectares and a further 30 Numbats were released there in 2011-2012. By December 2020, a third population was established within the largest fenced cat and fox-free area on mainland Australia, in Mallee Cliffs National Park. In partnership with NSW National Parks and Wildlife Service, 15 Numbats were released in 2020, and joined by another 31 individuals in 2021 and 2022. Numbat numbers are improving in parts of the lower Murray Darling region. The Yookamura Wildlife Sanctuary in South Australia has doubled its population to 42 (March 2024).
- Numbats are fascinating, endangered Australian **marsupials** known for their love of termites, their daytime activity, and unique parental care without a pouch. They are small, often called "banded anteaters" or "tiny tigers" due to their appearance, and are the sole **marsupial carnivore** that eats only termites, and can consume up to 20,000 termites in a single day! They use their excellent sense of smell to find termites and their long, sticky tongues to lick them up from logs and the ground, as they can't break into termite mounds themselves.
- Unlike many other **marsupials**, numbats are active during the day, which is called **diurnal**. This means they bask in the sun and search for food when termites are active, and then retreat into hollow logs, tree hollows, or **burrows** to sleep and avoid **predators** like foxes and **feral** cats.
- Numbats are a bit different from other **marsupials** because they don't have a pouch for their babies. Instead, their tiny young are born very small and attach themselves to their mother's teats, staying there for about six months before being placed in a **burrow** and suckled at night.
- They have long, pointed snouts to get into tiny spaces, and their fur is striped, helping them **camouflage** in their **environment**. They also have distinctive, brush-like tails.
- The numbat is the official emblem of Western Australia, which is where they are commonly found.

.... [1]

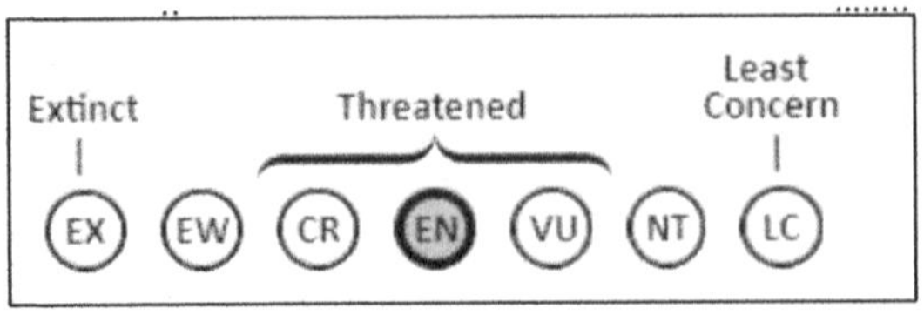

EPBC List = E = Endangered

Southern greater glider

Petauroides volans volans

- The southern greater glider is the largest gliding possum in eastern Australia. Its average body length is 35-45cm, and its long and furry tail measures 45-60cm. It weighs between 1 and 2kg.
- The southern greater glider could easily be mistaken for a fluffy tree-dwelling teddy bear, with its large, round ears and huggable, soft body. The thick fur increases its apparent size. The fur colour is white or cream below and varies from dark grey, **dusky** brown to light mottled grey and cream above.
- It has large ears with hair that projects past the outer edge of the ear.
- The southern greater glider occurs in eastern Australia, in eucalypt forests and woodlands, where it has a broad **distribution** from around Proserpine in Queensland, south through NSW and the Australian Capital Territory into Victoria
- The southern greater glider feeds exclusively on eucalypt leaves, buds, flowers and mistletoe.
- It shelters during the day in tree hollows and will use up to 18 hollows in its home **range**. They are recorded using hollows with a minimum diameter of 8cm. These hollows can help regulate its body temperature and keep it safe from **predators**.
- They occupy a relatively small home **range** with an average size of 1-3ha.
- Southern greater gliders give birth to a single young in late autumn or early winter, which remains in the pouch for approximately 4 months and is independent at 9 months of age.
- They are usually **solitary**, though mated pairs and offspring will share a den during the breeding season and until the young are independent.
- Very loyal to their **territory**.
- Southern greater gliders can glide up to a **horizontal** distance of 100m, including direction changes of as much as 90 degrees.
- Threats include:
    - Loss of **habitat**, **habitat fragmentation** due to human activities including farming, forestry, **urbanisation** and rural development.
    - Loss of hollow-bearing trees (which can take 300-400 years to develop hollows).
    - Alteration of microclimate conditions due to climate change, changes in rainfall, prolonged droughts, increased temperatures and heat waves. This **species** requires sufficient **moisture** on tree leaves to metabolise the toxin content.
    - Frequent or high severity fires.
    - Small population size making them **susceptible** to loss of individuals from predation and threats impacting **habitat** quality and food availability.
    - Barbed wire fences can entangle greater gliders and damage their gliding **membranes**.

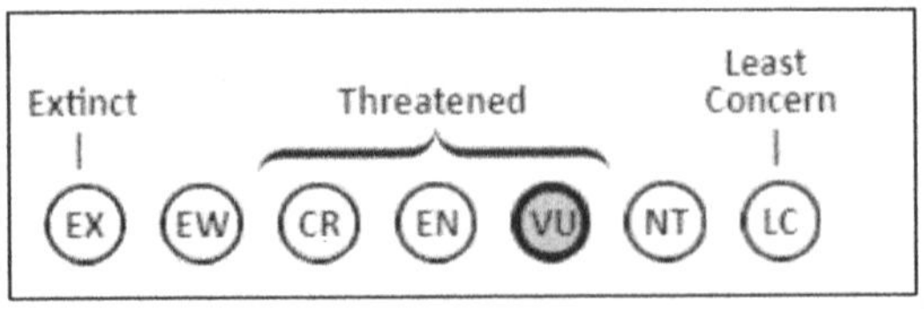

[1]

EPBC List = E = Endangered

Peron’s tree frog

Litoria peronii

- Peron's tree frogs are cool, **camouflaging amphibians** found in Australia with some **unique** features. They are known for their ability to rapidly change colour, their loud and distinct calls, and their interesting-looking eyes.
- They are a large tree frog species, reaching up to 75mm in total length.
- These frogs can change their colour quickly, from pale green-grey during the day to reddish-brown with emerald green spots at night. These colour changes are influenced by factors like temperature, time of day, and mood. This ability is particularly evident by the emerald green flecks on their backs. They can even change colour in less than an hour.
- Peron's tree frogs have unique bright yellow and black mottling on their armpits, groin, and the undersides of their thighs, which helps distinguish them from other frogs.
- Peron's tree frogs have a very unusual characteristic: their pupils are shaped like a cross and their iris is silver! This is uncommon in other tree frogs.
- Male Peron's tree frogs have a loud, repetitive call that sounds like a jackhammer or a "maniacal cackle," especially during their breeding season in spring and summer. They might even use drainpipes to make their calls louder!
- They inhabit a variety of **environments**, including forests, woodlands, shrublands, and even **urban** areas, often found near water sources like creeks, lagoons, dams, and ponds. They can also be found in tree hollows, fallen logs, and under bushes. They are good climbers and like places with lots of climbing material.
- Eggs are laid in small groups or singly and attached to twigs or **vegetation** in still water, with tadpoles taking around three to four months to develop. The tadpoles, are yellow and have stripes.
- They are **nocturnal**, meaning they are most active at night, **foraging** for insects. You might even see them on your window at night chasing moths!

[1]

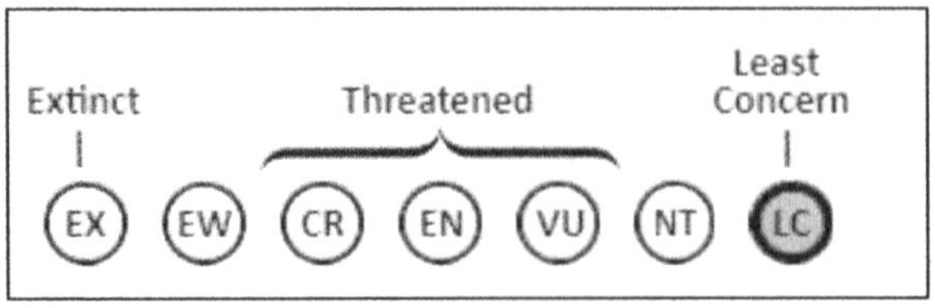

EPBC List = Not currently listed

Southern hairy-nosed wombat

Lasiorhinus latifrons

- The southern hairy-nosed wombat is the smallest of Australia's three wombat **species**. The southern hairy-nosed wombat can be found in South Australia, the southern parts of Western Australia, parts of Victoria, and some populations also exist in New South Wales.
- Southern hairy-nosed wombats can live for up to 30 years in the wild.
- They can grow to between 77-93cm in length, weighing between 19-32kg with an average of around 26kg, and are covered in soft, greyish fur.
- They are **solitary** animals and also **territorial**. While not always **aggressive**, they can use their powerful hind legs and teeth to **defend** themselves if they feel **threatened**.
- The southern hairy-nosed wombat can move fast, up to 40 km/h over a short distance.
- The southern hairy-nosed wombat builds large tunnel systems called warrens, some with many entrances, which can even be seen from space! These extensive **burrow** systems can be up to 3.5m underground.
- They are expert **burrowers**, spending much of their time underground and using their powerful forelimbs and long claws to excavate their **burrows**. They are primarily **nocturnal**, spending long periods in their **burrows**, especially during the hot summer months.
- Their pouches face backward to keep dirt out while digging, and they produce cube-shaped poo which is thought to help them mark their **territory** and prevent it from rolling away.
- The southern hairy-nosed wombat is an **herbivore** that eats grasses, roots, and bark. With their teeth continuously growing to deal with their tough diet.
- The southern hairy-nosed wombat is a **marsupial**, which means its babies develop in its mother's pouch.
- Koalas are the closest living relatives of wombats, sharing similar physical **characteristics** and also having backwards-facing pouches.

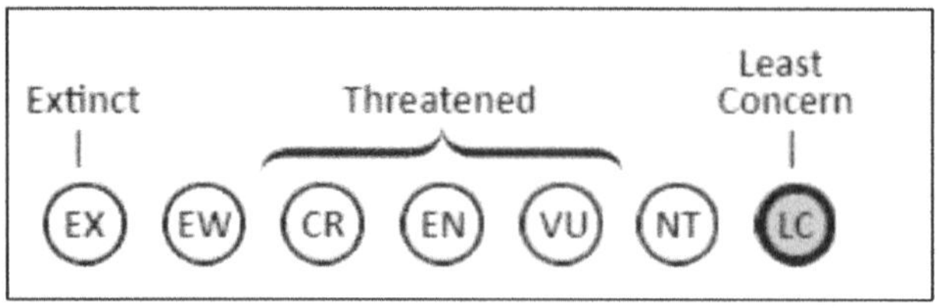

[1]

EPBC List = Not currently listed
Conservation status may vary by state.

Eastern long-eared bat

Nyctophilus corbeni

- The eastern long-eared bat, also known as Corben's long-eared bat or eastern tube-nosed bat, is found in eastern Australia, known for its distinctive long, tubular nostrils and yellow-spotted wings and ears, and it enjoys feasting on insects at night.
- The species is most prevalent in inland areas of the Great Dividing Range, with its range largely coinciding with the Murray Darling Basin, and the Pilliga Scrub region in New South Wales being a significant stronghold.
- They are relatively large bats, weighing around 14g, with a head and body length of 5-7cm, and their ears are larger than their head.
- This bat has a very unusual nose that sticks out with long, tube-like nostrils.
- You can easily spot them because their ears and the skin of their wings have bright yellow or yellow-green spots on them.
- They are **insectivores**, which means they eat insects like caterpillars, moths and beetles, usually hunting for food at night.
- **Foraging** occurs within a few kilometres of their roosting sites, generally at low height and around tree trunks or close to **vegetation**.
- These bats are great at using **echolocation** to find their food, making special high-pitched sounds that bounce off things to see in the dark, like a built-in sonar!.
- They like to live in different kinds of woodlands and rainforests. They **roost** in tree hollows, the hanging foliage of palms, in dense clumps of foliage of rainforest trees, under bark and in shallow depressions on trunks and branches, among epiphytes, in the roots of strangler figs, among dead fronds of tree ferns and less often in buildings.
- They are a medium-sized bat, with fur that is uniformly dark greyish-brown
- While they fly, they make a high-pitched "seep" sound, and they are slow but manoeuvrable flyers, good for catching insects in the air.
- Mating occurs in autumn with one or two young typically born in late spring to early summer.
- Known **predators** include goannas, hawks, owls, and **feral** cats. **Habitat loss** due to tree loss, modified fire regimes, and potential impacts from chemicals are significant threats to their survival.

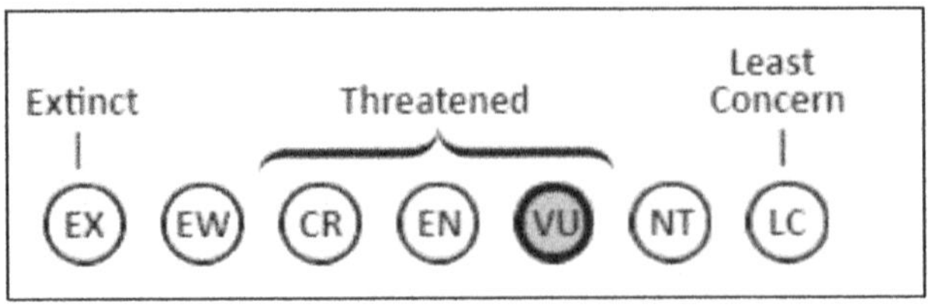

[1]

EPBC List = V = Vulnerable

Rainbow bee-eater

Merops ornatus

- The rainbow bee-eater is a spectacular bird. With its green, blue, chestnut and yellow **plumage**, slim build, long and slender curved **bill** and **distinctive streamers** extending from the end of its tail, it is simply beautiful. It has a golden crown and a red eye set in a wide black stripe from the base of the **bill** to the ears, which is edged with a thin blue line. The throat is orange-yellow, with a broad black band separating it from a green breast. The upper parts are green, with the flight feathers coppery and black-tipped. The underwings are bright orange with a black edge. The lower **abdomen** is blue. The tail, including the long **tail streamers**, is black with a blue tinge. Females have shorter, thicker **tail streamers** than males but are otherwise similar. Young birds are duller and greener, lacking the black band on the chest and the long **tail streamers**.
- There are many **species** of bee-eater in Africa, but the rainbow bee-eater is the only representative of the family in Australia. It is a native bird in New South Wales protected under the Biodiversity Conservation Act 2016.
- Rainbow bee-eaters are often heard before they are seen. They are known to be noisy and **conspicuous**. The rainbow bee-eater makes a series of loud, **melodious** "pir-r-r" calls, **characterised** by rapid vibrating and high pitch, usually in flight. A softer, slower call is exchanged between **perching** birds.
- Rainbow bee-eaters eat insects, mainly bees and wasps, but also dragonflies, beetles, butterflies, and moths. They catch flying insects on the wing and carry them back to a **perch** to beat them against it before swallowing them. Rainbow bee-eaters are **immune** to bee and wasp stings but will still rub them against the **perch** to remove the stings and **venom glands**. They can spot a potential meal from 50 metres away!
- Rainbow bee-eaters dig **burrows** in sandy banks/coffee rock to build their nests. The **burrows** can be over a metre long. While not **endangered**, the rainbow bee-eater's nesting period is **critical** to its survival. The **breeding** season for rainbow bee-eaters in NSW is from early November to January. Young birds leave the nest around February. Both parents **incubate** the eggs, sometimes with help from other bee-eaters in the **colony**.
- Bee-eaters have one of the most **complex social** systems of any bird. Many **species** (but not all) are **monogamous**, co-operative **breeders** living in large **colonies** of over 100 birds, which can reach 200 birds if conditions are favourable. Bee-eaters of both sexes alternate between being **breeders** and helpers in their **clan.**
- Threats to the rainbow bee-eater include **predation** by dingos, cane toads, monitor lizards, yellow-footed antechinus, and foxes. Dogs can dig up sandy nesting tunnels, and **human activity**, such as development and **quarrying**, disrupts their nesting.

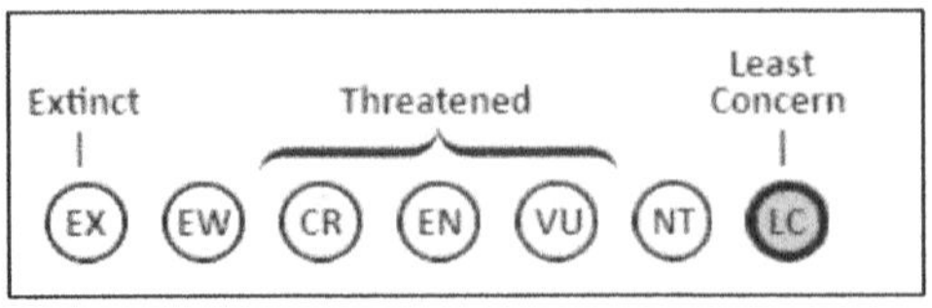

NCA List = not categorised
EPBC List = not categorised
Protected in NSW

Paradise parrot **EXTINCT**

Psephotellus pulcherrimus

- Sadly, this extraordinarily colourful bird, even by parrot standards, is the only mainland Australian bird recorded as extinct by the International Union for the Conservation of Nature (IUCN).
- The extinction of the Paradise parrot is likely to have resulted mainly from a combination of new pastoralist practices which **introduced** the burning by **graziers** to encourage fodder for their stock resulting in the loss of seasonal foods (at the critical time when the grass was in seed), **overgrazing** and land clearing, in combination with the effects of a severe drought in 1902, so that little amounts of seed were produced for several years. Hunting by bird collectors and predation by **introduced mammals** like cats and dogs may also be reasons.
- Its premature disappearance almost a century ago, meanwhile, remains prophetic today when it
- That the Paradise parrot – *Psephotellus pulcherimus* – was already on the verge of extinction by 1900 in its **habitat** on the Darling Downs region (part of the broader Murray-Darling Basin) in the Queensland **colony** speaks volumes about the dramatic **environmental** impact of colonisation on **native** grassy woodlands.
- The last live bird was seen in 1927. Extensive searches since the last live bird was seen have failed to produce any reliable evidence of its continued existence.
- The Paradise parrot was a medium-sized parrot **native** to the grassy woodlands extending across the Queensland and New South Wales border area of eastern Australia.
- It was once moderately common within its somewhat **restricted range**.
- The **plumage** was a mixture of turquoise, aqua, scarlet, black and brown.
- The parrots were small, measuring approximately 27-30cm in length.
- The tail was almost the same length as the body; unusual for a bird that, although a fast flyer, spent almost all of its time on the ground.
- The Paradise parrot mainly **inhabited** undulating river valleys, lightly timbered with eucalypt woodlands or open forests, often dominated by ironbarks and bloodwoods, with an **understorey** of **annual** and **perennial native** grasses; these areas were often dotted with **termitaria**.
- Though many Paradise parrots were taken into **captivity** in the 19th century, there are no current captive **populations** of this **species**, and none have been re**introduced** into the wild. There was a proposal to begin **captive breeding** of the **species** for release in the 1920s, but action occurred too late to save the **species**. Hybrid golden-shouldered parrots (*Psephotus chrysopterygius)* and Mulga parrot (*P. varius)*, which occur only in **captivity,** are said to appear similar to Paradise parrots but should not be mistaken for them.
- They were usually recorded in pairs or small **flocks**, probably family groups, of up to seven or eight, making their nests in hollowed-out termite mounds, often at or near ground level, and feeding, so far as is known, almost exclusively on grass seeds. Some observers state that **termitariums** were occupied by the same breeding pair in successive years. Less certain is confirmation of holes at riverbanks being tunnelled out by this **species**. A recorded **range** of one pair over a month was **restricted** to an area of two hectares. The nest sites were occasionally reported at **termitariums** previously occupied by the kingfisher *Todiramphus macleayii* or in **burrows** of a kookaburra *Dacelo leachii*, or at the base of a tree.

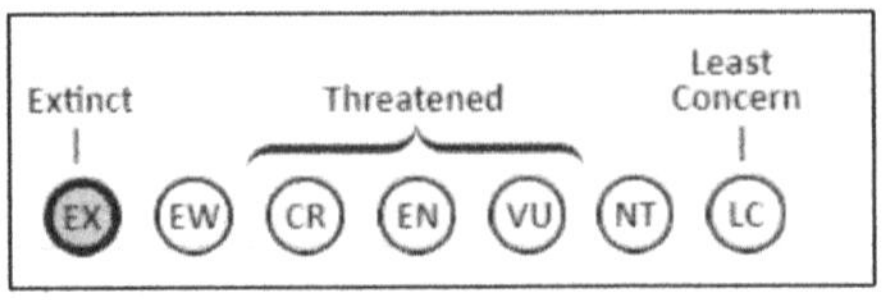

[1]

Just for fun: fins and feathers!

Bony fish anatomy:[37]

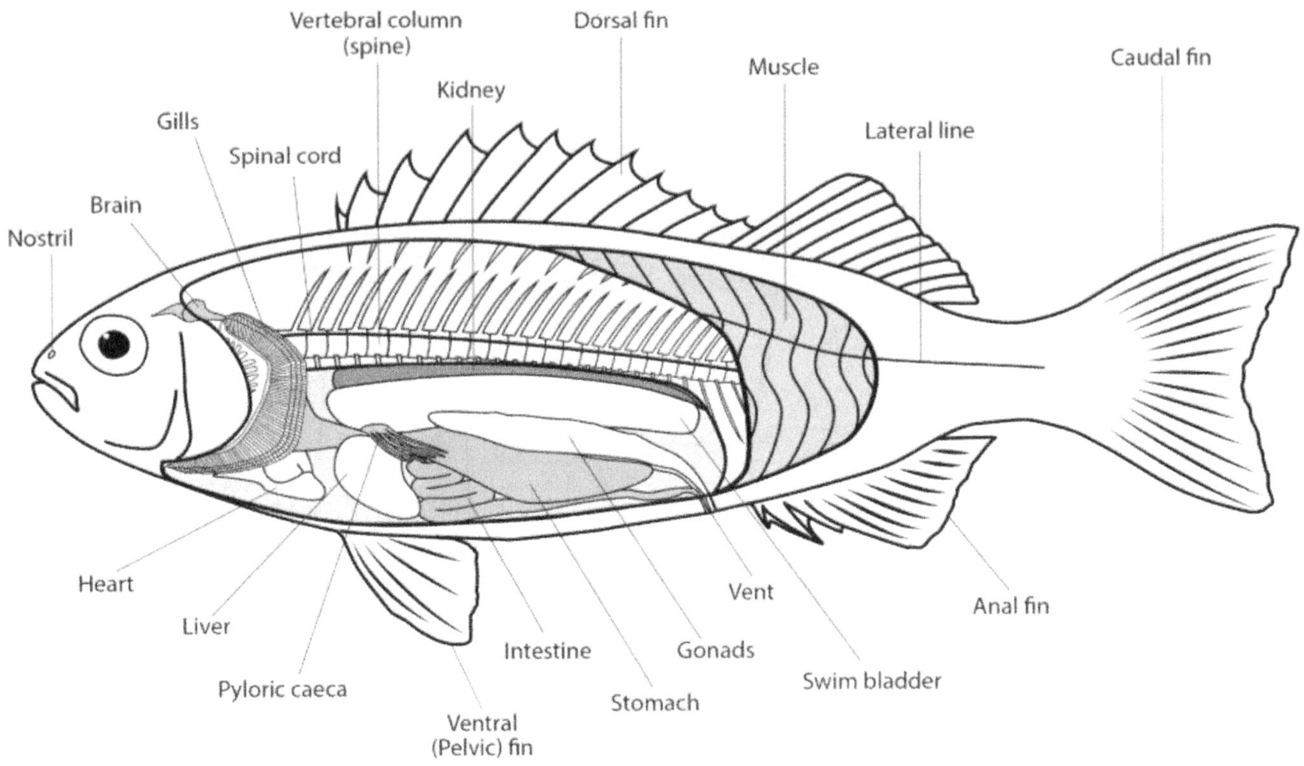

Basic bird anatomy:[38]

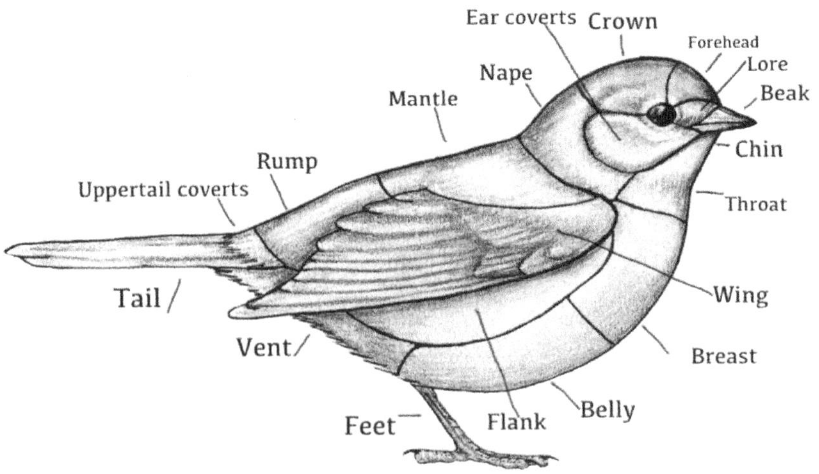

GLOSSARY – All those big or unusual words! ...

Some of these words are really unfamiliar, even to me as the author of this book, but I have included them because I think they could be useful for any person, child or adult, who has an interest in nature. For the most part, a synonym (word that means the same thing) has been used to try to keep these explanations as simple as possible. Definitions and synonyms from Dictionary.com and Thesaurus.com have been used on occasion. These definitions are not intended to be 100% perfect, but to give younger readers (and perhaps the adults reading with them) some other ideas to help them understand and explain what may be unfamiliar words, in the context of this book! Not all terms will be found in the text; however, they are still good terms to learn for any enthusiastic nature lover.

---

abdomen: belly, gut, midsection
aboriginal: original, earliest, native
aboriginal mythology: The "Dreaming" - the rich oral traditions, stories, and beliefs of indigenous Australians, passed down through generations
absorb: take in, soak up
abundant: ample, plenty of, extensive
abyssal (ocean): relating to or denoting the depths or bed of the ocean, especially between about 3000 and 6000m down
abyssopelagic (ocean zone): abyssal zone
achene (botany): a small, dry one-seeded fruit that does not open to release the seed
acid, acidic: substance like lemon, sour
acrobat, acrobatic/s: like gymnastics, athletic
adapt, adapted: change to suit, alter
adaptation/s: adjustment/s, alteration
adhere, adhesive: stick, sticky like glue
adjoin/s, adjoining: be next to and joined with
aggressive: ready or likely to attack or confront
agile: athletic, quick, energetic
agriculture, agricultural: the science or practice of farming
akin to: similar nature or character; related by blood
albino, albinism: pale or white colouring, usually with pink eyes
alga, algae: plants living in water
altitude, altitudes: height in relation to sea or ground level
ambush: trick, trap, a surprise attack
ambush hunter, ambush predator: an animal that prefers to capture or trap prey by stealth or strategy
amphibian: an animal that can live both on land and in water, cold-blooded vertebrate animal
amphidromous: organisms, typically fish, that regularly migrate between freshwater and saltwater environments, but not for breeding purposes
anal: related to or near the anus
analysis: study, test, investigate
anatomy/anatomical: structure of a plant or animal, the way it is built
angler: an animal/person who catches fish
annual, annually: yearly, once a year
antenna, antennae: feelers
antennules: small antenna, especially one of the foremost pair of a crustacean
anterior (anatomy): towards the front
anther: the part of the stamen of a flower that produces and contains pollen and is usually borne on a stalk
antivenom: a substance used to neutralise or counteract the effects of a venom
anus: the opening at the end of the digestive system

apex predator: a meat eating animal that preys on other animals but has no natural predators of its own, at the top of the food chain

aquarium, aquaria: fishbowl, fish tank

aquatic: occurring in water

aquatic vegetation: plants that rely on wet environments

arboreal: living in or among trees

archaea: microorganisms which are similar to **bacteria** in size and simplicity of structure but radically different in molecular organisation. They are now believed to constitute an ancient group which is intermediate between the **bacteria** and **eukaryotes**

arid (of land or a climate): having little or no rain; too dry or barren to support vegetation

arthropod: an animal with a segmented body, jointed limbs and usually a shell including insects, spiders, crustaceans

artificial: not natural, made by humans

assimilate: ingest/digest, soak up

Austral spring: a season of spring in the Southern hemisphere where the greatest amount of ozone is lost, generally beginning in September

awn: a stiff bristle, especially one of those growing from the ear or flower of barley, rye, and many grasses

axis: an imaginary line that goes through the middle of something

bacteria/bacterium: a member of a large group of unicellular microorganisms which have cell walls but lack organelles and an organised nucleus, including some that can cause disease.

bacterivore/bacterivorous: an organism which obtains energy and nutrients primarily or entirely from the consumption of bacteria.

barb: a sharp projection near the end of an arrow, fish hook, or similar object, which is angled away from the main point so as to make extraction difficult; a fleshy filament at the mouth of some fish such as barbel and catfish

barbed: having a barb or barbs

barbel: a fleshy filament growing at the mouth of some fish

bathypelagic (ocean zone): midnight zone

bill (of birds): beak; the hard, pointy part of a bird's mouth

billabong/s: small lake usually formed when the path of a creek or river changes leaving a dead end

biocide/s: a substance that kills life

biodiversity: differences in biology within plant and animal species

bite force: force applied between the upper and lower mouth structures, jaw and/or teeth

bivalve: any mollusc with two shells hinged together

botany: the scientific study of the physiology, structure, genetics, ecology, distribution, classification, and economic importance of plants

bracken: type of big, bushy, green fern

brackish: water that is saltier than fresh water but not as salty as seawater

brood: a number of offspring, clutch, youngsters

brood pouch: a pouch where eggs develop and hatch

bunyip: the bunyip is a creature from the aboriginal mythology of southeastern Australia, said to lurk swamps, billabongs, creeks, riverbeds, and waterholes

burrow: hole or tunnel dug by an animal, usually for a home or hiding place

buttress roots: large, wide roots sometimes known as plan roots on all sides of a tree

calligraphy: decorative handwriting or handwritten lettering

camouflage: disguise, mask, concealment

canine/s: a pointed tooth between the incisors and premolars of a mammal, often greatly enlarged in carnivores

canopy: the upper part of a tree or plant group that's made up of leaves and branches. It's also known as the overstory

captive breeding: bred in captivity

captivity: kept in a place like a zoo or a cage

carapace: a bony or chitinous shell, test or shield covering some or all of the dorsal part of an animal, e.g. turtle

carnivore/carnivorous: meat eating

carotenoid: nature's food colouring! They're the bright yellow, orange, and red pigments that give fruits and vegetables their colourful appearance

carpal: relating to the bones forming the wrist

carrion: dead and rotting flesh of an animal

cartilage: strong, bendy, and rubbery tissue

cartilaginous tissue: made out of cartilage like our ears and nose tips

caudal (anatomy): of, relating to, or being a tail; directed toward or situated in or near the tail or posterior part of the body

cauline (botany): relating to or growing from a plant stem, particularly the upper part of the stem

characteristic: distinctive, singular feature

chemical cues: hints that are chemical, could be particular smell

chenopod grassland: *Chenopodioideae* are annual or perennial herbs, subshrubs, shrub or small trees

chromatophore: tiny colour-changing cells under an animal's skin

chrysalis: the hard-shelled pupa of a moth or butterfly

chytrid fungus: an infectious fungal disease which can be fatal to amphibians

chytridiomycosis: is an infectious disease of amphibians – see **chytrid fungus**

circulate: move in a circle, pass from place to place, spread

circumference: the distance/measurement all the way around an object/circle

cite, cited: quote, name, reference

classify, classification (scientific): the sorting of things into groups based on how they are alike

cloaca/cloacal: a shared exit for bodily waste and reproductive function

clutch (of eggs): total number of eggs/young in one nesting

coastal dredging: excavating material from a coastal water environment

co-exist: live or exist together

colony/colonies: group, settlement, swarm

colour morph: a specific body colouration variation

combatant/s: fighter, battler

commercial: merchant, business, of trade, public

commercial port: a place for ships to load or unload things for buying and selling

complex: complicated, convoluted

cones (eyes): specialised cells that allow us to see colour and fine details, particularly in bright light

configuration: composition, shape, structure

congregate: gather into a crowd or mass

conservation: saving and protecting

conspicuous: easily seen or noticed; standing out from a group

construct: build or put together

consume: eat, drink or use something up
contamination: infection, pollution, corruption
continent: big, solid area of land – Earth has seven continents
co-operative: working with others to help make things happen
coronal (anatomy): dividing body into front and back
counterpart: equals, fellow, like, twin
courting, courtship: set of display behaviours in which an animal, usually a male, attempts to attract a mate
cranial (anatomy): towards the head
creature: a living animal or living insect
creche: a group of young cared for communally
credible: believable, reasonable, likely
creek: small stream or river
crepuscular: resembling or relating to twilight
crevice: slit, crack, cleft, gap
critical: very important
critically endangered: facing an extremely high risk of extinction in the wild
critter: informal word for any living creature
crustacean/s: mainly aquatic arthropod typically with body covered with a hard shell or crust, such as lobsters, shrimps, crabs, barnacles, wood lice
cryptic: secret, mysterious, hidden
ctenoid scales: scales with spines or comb-like teeth along their free edge
cull/culling: reduce the population of (a wild animal) by selective slaughter
culprits: wrongdoer, offenders
cultivate: breed, raise, tend to, plant
cultural significance: of some value for past, present or future generations
culvert/s: tunnel carrying a stream or open drain under a road or railway
currency: money, legal tender
cyanobacteria: formerly known as blue-green algae, are photosynthetic microscopic organisms that are technically bacteria
cylinder, cylindrical: 3D shape like a tube with circles on either end
debris: scattered pieces of rubbish or remains
decade: 10 years
decline/declining: drop, gall, decay, weakening
decorticate/decorticating: remove the bark, rind, or husk from.
deep (anatomy): further from the surface
defence mechanism/s: adaptations that help avoid predators or mitigate adverse effects.
defense, defensive: protecting or defending against something
deforestation: the action of clearing a wide area of trees
degrade, degradation: reduce in quality, lessen, weakening, deterioration
dense: thick, heavy, impenetrable, solid
derivation (words): where the word/s came from
destruction: breaking something apart, causing so much damage it can't be used or fixed, or even removed
detritus: small pieces of disintegrated material such as organic matter
difference between endemic and native? An endemic species is a native species but found only in a particular area, large or small
digest: to break down food into smaller pieces to use for energy and growth
digestive system: the parts of the body that work together to turn food and liquids into the building blocks and fuel that the body needs for energy and growth
digit/s: finger, thumb or toe
dimorphic: two different forms
dioecious: having the male and female reproductive organs in separate individuals
disease: illness or sickness characterised by specific signs or symptoms
dissipate: disappear, waste or fritter away

distal (anatomy): further from the centre
distinction/s: noticing the difference between two things
distinguish, distinguishes: mark off as different
distinctive, distinctively: having a special quality, style, colour, etcetera
distribute, distribution: divide into, disperse through a space or over an area, spread, scatter
diurnal: of or during the day.
diverse/diversity: different or having variety
dominant: more strong than others
dorsal, dorsum (anatomy): of, relating to, or situated at the back, or dorsum of
drainage: removing excess water from an area
drupe: a type of fleshy fruit that contains a single hard seed enclosed in a stony layer
duck blind: shelter, often camouflaged with reeds and grasses, for hidden observation
dusk: time of day when the sun is setting and it's starting to get dark, just before night
dusky (of colour): of a darker colour, shadowy
dwindle: become less or smaller gradually
echolocation/echolocate: the location of objects by reflected sound
ecosystem: a geographic area where plants, animals, and other organisms, as well as weather and landscapes, work together to form a bubble of life
ectomycorrhizal: a symbiotic association of fungi with the feeder roots of higher plants
edible: fit to be eaten as food, eatable
eel: long snake-like fish
elaborate: complicated, detailed, **complex**
elliptic (botany): the shape of structures like leaves, being widest at or near the middle and tapering to both ends, similar to an oval
elusive: tricky, difficult to catch, shy
endemic: native species found only in a particular area
embryo: baby in its very early stages, before it's born; animal or plant that's just starting to grow and develop
endangered: there is a very high risk of extinction due to rapid population declines of 50 to more than 70% over the previous 10 years, a current population size of fewer than 250 individuals
environment, environmental: the natural surroundings that make up the world we live in
EPBC: Environment Protection and Biodiversity Conservation Act 1999 (EPBC Act) is Australia's main environmental law
ephemeral: lasting for a very short time; fleeting or transitory
epicormic buds: dormant buds that lie beneath the bark of a tree
epipelagic (ocean zone): sunlight zone
estimate, estimated, estimating: guess, projection
estuary/estuarine: an area where a freshwater river or stream meets the ocean
eukaryotes: any cell or organism that possesses a clearly defined **nucleus**
European settlement: 1788-1810 early European settlement
evade: avoid or get away from something, often by being clever or sneaky
evolve, evolved: grow, mature, advance
evolution, evolutionary: change, expand, progress, transform
excavate: dig out, burrow, uncover
excretion/s: discharge by excreting
exotic: of foreign origin or character, introduced from abroad
extensive, extensively: covering a large area
extinct/extinction: having died out, no longer existing

extreme weather events: occurrences of unusually severe weather or climate conditions that can cause devastating impacts on communities and agricultural and natural ecosystems

exudate: substance coming out, ooze

fatal: causing death

faunal region: any of six or seven areas of the world defined by animal geographers on the basis of their distinctive animal life

feral: wild, having reverted to a wild state

fertilised, fertilisation: make productive, capable of development

fire regimen: pattern, frequency, and intensity of the bushfires and wildfires that prevail in an area over long periods of time

flank/s: side between ribs and hips

flock: group of animals, especially birds like geese or chickens, or sheep and goats, that stay together

follicle/follicles: dry fruit formed from a single carpel that splits open along one side to release its seeds

forage, foraging: look for food

formidable: something that is really strong, impressive, or even a little scary, but also something you might respect

fossil: the preserved remains or traces of a living thing that lived long ago

fragment, fragmented, fragmentation: broken off, scattered, isolated, separated

frugivore/frugivorous: an animal that thrives mostly on raw fruits or succulent fruit-like produce of plants such as roots, shoots, nuts and seeds

fungus, fungi, fungal: simple organism, or living thing, that is neither a plant nor an animal

gangly: tall and thin and moving with a loose-jointed awkwardness

generalist feeder: feeds on a wide variety of things and thrive in various environments

generic: referring to all members of a genus

gene/s: the basic unit of heredity. This means that genes determine what traits are passed down from a female and male to their offspring. Genes exist inside the cells that make up living things

genetic/s: traits carried in genes, which are like special instructions for making a body

genus: the major division of a family, usually of more than one species

germinate/germination: when a seed starts to grow into a plant

gestation: the process or period of developing inside the womb between conception and birth

glabrous: smooth and hairless

global/globally: worldwide, comprehensive

globose: having the shape of a globe or sphere; round

glossy: shiny, lustrous, polished

graffiti: writing or drawings, scribbled, scratched, or sprayed, usually illicitly

granivore/granivorous: an animal that primarily feeds on seeds and grains

granular: something that is made of tiny grains or particles, like sand or sugar, or that looks like or feels like it is made of those small bits

graze, grazing: eating, feeding, nibbling, munching

groundwater: water below surface of ground

habitat/s: place where usually found, natural environment of an organism

harem: a group of female animals sharing a single mate

harvest, harvested: gather food

hatch, hatching: baby animal coming out of its egg, or that an egg is breaking open to release the baby
hatchling: young bird, reptile or fish just hatched from egg
heath/heathland: an area of open uncultivated land, typically on acid sandy soil, with characteristic vegetation of low-growing plants, shrubs and grasses.
herbicide/s: a substance that is **toxic** to plants
herbivore/herbivorous: an animal that primarily eats plants
herper/s: a person who is interested in reptiles and amphibians, or herps
hibernate, hibernation: spend the winter in a dormant state
hinder: make harder or slower to do, make difficult
honeydew: sugary material secreted by some insects such as aphids, scale
horizontal: parallel to horizon
host plants, host trees: plants that an organism lives on and lives off (parasitic or semi-parasitic plants)
hostile: aggressive, warlike
human disturbance: people disturbing the environment: for example - tree removal. road building construction near shorelines and streambanks
humid, humidity: moisture or water vapour in air
hydrology: the branch of science concerned with the properties of the earth's water, and especially its movement in relation to land.
iconic: representative of
instar: a developmental stage between moults, or shedding of the exoskeleton
illegal: against the law, not legal
immature: something that is not fully developed or grown
immune to: not being affected by something
impoundment (of streams): damming, enclosing
incisor/s: a narrow-edged tooth at the front of the mouth, adapted for cutting
incubate, incubated, incubation: brooding
indigenous: having originated in and being produced, growing, living, or occurring naturally in a particular region or environment
indigenous people: First Nations people
inferior (anatomy): below, lower than
inflorescences: the complete flower head of a plant including stems, stalks, bracts, and flowers
inhabit, inhabits: live in, occupy, populate, reside
insecticide: an agent (usually chemical) that destroys insects
insectivore/insectivorous: a carnivorous animal or plant which eats insects
intertidal: between high and low water marks
intimidate/d: scare, frighten, bully, terrify
introduce/introduced (of species): plant or animal that wasn't originally there but was brought in by humans, either on purpose or by accident
invasive: threatening a plant or animal community, by spreading prolifically
invertebrate/s: animals without backbones
iridescent: something that changes colours when you look at it from different angles
iridophores: an iridescent chromatophore
iris (of eye): the coloured tissue at the front of the eye that contains the pupil in the centre
isolation, isolated, isolate: detach, separate, place apart
IUCN: International Union for the Conservation of Nature
joey/s: young marsupial animal
juvenile/s: youth, junior, infant, young
keystone species: an organism that has a disproportionately large impact on its ecosystem relative to its abundance

kinetoplast, kinetoplastids: a group of flagellated protozoans that are distinguished by the presence of a DNA-containing region, known as a "**kinetoplast**", in their single large mitochondrion

labellum: a central petal at the base of an orchid flower, typically larger than the other petals and of a different shape

lactate, lactating: producing and secreting milk from the mammary glands

lantana thicket: compact shrub which can grow in dense thickets, areas of trees and bushes growing close together

larva, larvae: immature, wingless, feeding stage of an insect that undergoes complete metamorphosis

lateral (anatomy): towards the side

legume (botany): a dry fruit (a pod) that typically splits open on two sides at maturity, and it is produced by plants belonging to the Fabaceae family (also known as Leguminosae), which includes well-known edible species like beans, peas, and lentils

lichen: a complex, plantlike organism formed by a symbiotic relationship between a fungus and a photosynthetic partner, which can be an alga or a cyanobacterium

life cycle: the series of changes in the life of an organism

lifespan: the period of time of existence from birth to death or the end of useful function

lignotuber: a woody swelling at the base of a plant that contains dormant buds, nutrients, and carbohydrates

ligule (botany): a narrow strap-shaped part of a plant, especially a membranous scale on the inner side of the leaf sheath at its junction with the blade in most grasses and sedges

litter: number of young born in one birth

littoral: relating to or situated on the shore of the sea or a lake

littoral rainforest: a closed forest community that grows near the ocean, usually within 2km

local, locally, localised: restricted to one place

locally extinct: no longer present in a specific geographic region but still exists elsewhere

longitudinal/longitudinally: running lengthwise rather than across.

lore: the region between the eyes and nostrils of birds, reptiles, and amphibians

macro- (prefix): large; large-scale.

mammal: warm-blooded vertebrate having hair or fur, producing milk for the young and giving birth to live young – exceptions are the **monotremes.**

Māori people: indigenous Polynesian people of mainland New Zealand

marine: of or relating to the sea

marsh/marshes/marshland: type of wetland ecosystem characterized by poorly drained mineral soils and by plant life dominated by grasses

marsupial: animals in the order *Marsupialia* which have a pouch for their young

marsupium: pouch found in female marsupials

maternity: the period during pregnancy and shortly after childbirth

maternity colonies: group of pregnant or lactating female bats that gather together in a roost to give birth, nurse, and raise their young

mature, maturity: fully developed, having completed natural growth and development
mechanism: the fundamental processes involved in or responsible for an action, reaction, or other natural phenomenon
medial (anatomy): towards the middle
melody/melodious: sounding pretty and musical
membrane/s: sheath, sheet, film of tissue
mesial: being or located in the middle or a median part
mesopelagic (ocean zone): twilight zone
metabolism: the chemical reactions in cells that change food into energy
metamorphosis: change in life cycle e.g. pupa to caterpillar to butterfly
methodical: orderly, structured, deliberate
micro- (prefix): very small, tiny
microalgae: microscopic algae invisible to the naked eye; phytoplankton typically found in freshwater and marine systems, living in both the water column and sediment. They are unicellular species which exist individually, or in chains or groups
microendemic: species with very small geographic distributions (ranges)
microorganism: an organism that is so small it can only be viewed using a microscope. Also called microbes, these tiny living things are found everywhere in water, soil, air, and the human body, with examples including bacteria, fungi, and protozoa
microphyte: a microscopic plant
microscopic: incredibly small, unable to be seen without a microscope
migrant: drifter, traveller, wanderer
migrate, migration, migrating: going from one country, region or place to another
migratory: characterised by migration
mimic, mimicking, mimickry: copy, simulate, imitate
mitochondria, mitochondrium: the tiny powerhouses inside our cells! They break down food we eat to give us energy, so we can move, think, and do everything else we do...think of them as little energy factories
moist: slightly wet, damp
moisture: the presence of a liquid, especially water, often in small amounts
molar: the back teeth in the mouth, shaped like flat, grinding surfaces
mollusc/s: animal that has a soft body, no spine, and is often covered with a shell
monitor: observe, watch, keep an eye on
monoecious: having both the male and female reproductive organs in the same individual
monogamous: having only one mate at a time
monopolise: have complete and sole control over, take over control
monotreme/s: mammals who lay eggs and have no teats but secrete milk to feed their young
montane (geography) : specific ecological zones and ecosystems found at high elevations on mountains, particularly those below the timberline
morph (biology): one of the distinct forms or variations found within a species, especially when these variations are visually or behaviourally different
morphology, morphological: form of a thing
motivate. motivated: inspire, persuade, prompt, spark
moult: shed, cast, peel - when an animal sheds its old outer covering (skin/feathers, for example) to make way for a new one or new ones

myth, mythology: belief, legend, lore, tradition
native: an animal or plant that has lived in a specific area for a long time, naturally, without being brought in from somewhere else
navigate, navigation: find a route to a destination
navigator: a person or animal who navigates
NCA: Nature Conservation Act 1992
nectar: sweet, sticky liquid made by flowers
nectarivore, nectarivorous: an animal which derives its energy and nutrient requirements from a diet consisting mainly or exclusively of the sugar-rich nectar produced by flowering plants
nocturnal: active at night, relating to night
nomadic: wandering, travelling, roving
no-take: not allowed to be taken/caught
nucleus/nuclei: the structure in a cell that contains the chromosomes
nutrient/s: source of nourishment, food
nutrient-rich: rich in nutrients, nourishing food
nymph (insect): young insect that looks a bit like the adult, but smaller and without wings
oblanceolate (botany): having a broad rounded apex and a tapering base. simple, unsubdivided. (botany) of leaf shapes; of leaves having no divisions or subdivisions
obovate (botany): egg-shaped with the narrower end at the base. simple, unsubdivided. (botany) of leaf shapes; of leaves having no divisions or subdivisions
oceanic: marine, of the oceans
oesophagus: a muscular tube in the digestive system that carries food and liquids from the pharynx (throat) to the stomach
Olongburra: This specific name refers to the aboriginal people, the Olongburra, who once lived in the northern half of K'gari (Fraser Island)
omnivore, omnivorous: eating both animal and plant foods
opportunistic: taking opportunities when they present
opposable, opposing: able to be placed opposite to something else
organelle: a **subcellular** structure that has one or more specific jobs to perform in the cell, much like an organ does in the body. Among the more important cell **organelles** are the **nuclei**, which store **genetic** information; **mitochondria**, which produce chemical energy; and **ribosomes**, which assemble proteins
organic: live, living, natural
organic material: matter that is made of organic compounds from the remains of organisms such as plants and animals, and their waste products
organism: animal, plant, structure, creature
osmeterium: a defensive, eversible (protrusible) organ found in swallowtail butterfly caterpillars, located just behind their heads, which they use to ward off predators by emitting a foul-smelling fluid and displaying bright, often red or orange, "horns"
ovary: the organ in a female animal that produces eggs and certain hormones; the part of a flower that contains the seeds and grows into a fruit
overexploit, overexploitation: overuse, excessive use of resources
overfish, overfished, overfishing: catching too many fish at once so that the breeding population is too reduced to be able to recover

overhunt, overhunted, overhunting: the act of hunting animals or birds to an excessive degree, which can be harmful to the population or the environment

overgraze, overgrazed, overgrazing: allowing grazing animals to reduce grazing land to beyond the point of recovery

ovum: an egg cell

ovule: an outgrowth of the ovary of a seed plant that after fertilization develops into a seed

paralysis: inability to move, can sometimes affect entire body systems such as breathing, movement, heart beat

parasite, parasitic: an organism that lives on or in an organism of another species (the host) from the body of which it obtains nutriment

passerine: relating to or denoting birds of a large order distinguished by having feet that are adapted for perching, including all songbirds

pastural/pastoralism: something related to rural life, especially the life of shepherds or country life

patagium/patagia *(pl.)***:** wing membrane (bat), extensible fold of skin (insects, gliding mammals and reptiles), small process on the anterior thorax seen especially among butterflies and moths

pectoral fin: see fish anatomy diagram

peduncle (botany): the stalk of a flower

pelvic: having to do with the pelvis

pelvices: plural of pelvis

pelvic fin: see fish anatomy diagram

pelvis: the bony structure of the body below the abdomen that is located between the hip bones and contains the bladder and rectum

perch, perching: place where a bird can sit and rest, like a branch or a stick

perennial: lasting or existing for a long or apparently infinite time; (of a plant) living for several years

perianth (botany): the outermost, non-reproductive part of a flower, consisting of the calyx (sepals)and corolla (petals) when they are distinct, or tepals when they are undifferentiated

perimeter/s: boundary, edge, borderline

pesticide/s: a substance used for destroying insects or other organisms harmful to cultivated plants or to animals

petiole (botany): also known as a leafstalk, is the stalk that connects a leaf's blade to the stem of a plant

photosynthesis: how plants make their own food using sunlight, water, and air to produce oxygen and sugar, which they use for energy

phyllode (botany): modified petiole or leaf stem, which is leaf-like in appearance and function

phytoplankton: consisting of microscopic plants

piscivore/piscivorous: a carnivorous animal that primarily eats fish

pistil: the female reproductive part of a flower

placid: gentle, peaceful, quiet

plankton, planktonic: floating, drifting, or somewhat motile organisms occurring in a body of water, primarily comprising microscopic algae and protozoa

plastron: the ventral part of a turtle shell turtle

plume/plumage: feather (singular), feathers of a bird, collectively

poison, poisonous: likely to cause serious harm or death

pollen: a fine powdery substance, typically yellow, consisting of microscopic grains discharged from the male part of a flower or from a male cone. Each grain contains a male gamete that can fertilise the female ovule, to which pollen is transported by the wind, insects, or other animals

pollination: the act of transferring pollen grains from the male anther of a flower to the female stigma to help make new plants

pollinator, pollinators: animal, insect or other agent that transfers pollen to a plant allowing fertilisation

pollution: the introduction of harmful materials into the environment

polychaete worm: poly means 'many', chaete means 'hair', so these are a species of marine worm, commonly known as bristleworms.

polyclad, polyclads: any free-swimming, marine flatworm of the order Polycladida, having a broad, flat body and a many-branched gastrovascular cavity

polyoestrous: can raise several litters a year

population: number of inhabitants of an area

posterior (anatomy): behind, towards the back

precarious: dependent on circumstance, unsure, indeterminate, unreliable, doubtful

precipitation: what happens when water in the sky (clouds) falls back down to Earth as rain, snow, sleet, or hail

precocial: (of a young bird or other animal) hatched or born in an advanced state and able to feed itself and move independently almost immediately

predator/s, predation: an organism that exists by preying on other organisms

predict, prediction: to be able to tell in advance

pregnant: containing a developing embryo, foetus, or unborn offspring within the body

prehensile: adapted for seizing, grasping, or taking hold of something

premature: happening or done too soon, or before the expected or natural time

prey, preying: an animal, insect or organism hunted or taken for food, especially by a carnivorous animal

primary: most important, main

prokaryotes: organisms whose cells lack a **nucleus** and other **organelles**. **Prokaryotes** are divided into two distinct groups: the **bacteria** and the **archaea**, which scientists believe have unique evolutionary lineages. Most **prokaryotes** are small, single-celled organisms that have a relatively simple structure

probe/probing: examining with the beak

prominent: standing out, obvious

protozoa, protozoan: any of a diverse group of eukaryotes that are primarily unicellular, existing singly or aggregating in colonies, are usually non-photosynthetic and often classified further into phyla according to their capacity for means of motility

protuberance/s: a rounded, outward projection or swelling on a surface

proximal (anatomy): closer to (root of the limb)

puggle: baby echidna or platypus

pulmonate, pulmonates: having lungs or lunglike organs

pupa, pupae *(Pl.)*: an insect in the nonfeeding, usually immobile, transformation stage between the larva and the adult stage

pupil (eye): the black opening in the middle of the coloured part of your eye

pursue: chase, track, follow, trail, hunt

pursuit: chase, hunt, quest, run

quarry (prey): something being hunted or pursued, specifically prey
quarry, quarrying: an open excavation usually for obtaining building stone, slate, or limestone
radio-tracking: a way of sending sounds or other information through the air from a transmitter attached to the animal to enable researchers to track it
rainforest canopy: the denser layer of greenery about 20 feet thick formed by the next tallest trees.
range (of species): an area where a member of a particular species can be found during its lifetime
rapid/rapidly: quickly, quick
raptor: usually a bird or prey, such as an eagle or a hawk
rare: not common or usual
recognise, recognition: seeing something you have seen before or can clearly identify by other means
recreation, recreational: fun activity, amusement, hobby
refuge: place that offers safety and protection from danger, trouble, or bad weather
regenerate: the process of restoring damaged or missing cells, tissues, organs, or even entire parts
remnant: remains, fragment, piece
reproduction/reproduce/reproductive: the production of offspring
reptile: cold-blooded vertebrate
residential development: housing development
respiratory system: breathing system, lungs
restrict, restricted: confine, restrain, regulate, limit
retina, retinae *(Pl.)***:** inside back of eye of several layers
revere, revered: respect or admire greatly
reverse: backward, contrasting, opposite, contrary, change
reversible (physiological): something that can be changed from back to front or front to back
rhomboid, rhomboidal: parallelogram in which adjacent sides are not equal
ribosome: an intercellular structure made of both RNA and protein, and it is the site of protein synthesis in the cell
riparian: situated on the bank of water
robust: strong and healthy
rodent/s: a gnawing mammal of an order that includes rats, mice, squirrels, hamsters, porcupines, and their relatives, distinguished by strong constantly growing **incisors** and no **canine** teeth
rodenticide: a type of poison used to kill or control rodents
rods (eyes): specialised cells which are sensitive photoreceptors to light and help us see in low-light conditions. They are primarily responsible for night vision and peripheral vision, but they don't perceive colour
roost, roosting: perch where birds rest or sit
rufous: tinged with red, brownish red
rump: hindquarters, buttocks, sacral region of a bird
rural: relating to the countryside and not to towns or cities
sagittal (anatomy): dividing body into right and left
saline: containing or impregnated with salt
salinity: saltiness
samara: a dry indehiscent usually one-seeded winged fruit
satellite: an artificial body placed in orbit round the earth or moon or another planet in order to collect information or for communication.
satellite tracking: using satellites orbiting earth to track

satellite transmitter: device which sends a signal to a satellite
saturate, saturated: soaked, soggy, waterlogged
savannah: flat grassland, plain
scarce, scarcity: not plentiful, hard to find or hard to get
school, schooling (of fish): group of fish moving together
sclerophyll forests : vegetation that has adapted to long periods of dry, hot weather
sea squirt: invertebrate, usually fixed potato-shaped organism using sac-like filter feeding
secretive: keeping secrets
sedentary: staying in one place, not moving much
sedge: plant that looks like grass, but it grows in wet areas
sediment, sedimentation: organic matter settling to bottom of a liquid
sedimentary: of or relating to rocks formed by the deposition of sediment
segment/s: part of, divided parts
semi- (prefix): half, partially, incompletely
semi-parasitic: plant which gets water or food from host but can also produce its own food
semi-prehensile: able to grasp but cannot really hang on tightly
semi-transparent: almost see through
sepal: the green, leaf-like part of a flower that protects the bud before it blooms and supports the petals when the flower is open
serrated: having a jagged or toothed edge, like a saw
settlement: a place, typically one which has previously been uninhabited, where people establish a community
shellfish: seafood like molluscs, crustaceans
sheoak: type of tree
sibling: each of two or more children or offspring having one or both parents in common; a brother or sister
sift: an act of sifting something, especially so as to isolate that which is most important
significant, significance: very important or has a special meaning
silt: fine sand, clay or other material carried by running water
siltation: water pollution by silt
skim, skimming: to pass lightly or hastily; glide or skip along, above, or near a surface
snout: the long nose and mouth of an animal that sticks out
snout-to-vent: a measurement of the body length of an animal, specifically from the tip of the snout to the cloacal opening, or vent
social: friendly, like company
solitary: by itself, single
spawn, spawning: reproduce, generate (usually fish or frog eggs)
sparse: something that is thinly spread out or not dense
species: one of the classes of things included with other classes in a genus
specific: clear-cut, particular, precise, unique
specimen: sample, example
spinifex: grass plant with stiff, sharp-pointed leaves that grow in dense masses
spinose: having many sharp, pointed things sticking out; spiny, prickly
spire: a part of the coiled shell of molluscs. The spire consists of all of the whorls except for the body whorl
spoil heap: an untidy pile or mass usually of refuse material from an excavation

spore: a reproductive cell that can develop into a new individual without uniting with another reproductive cell

stalling: stopping (flying)

stance: way of standing physically or figuratively

stamen: the male part of a flower that makes pollen

stigma/s: the part of a flower that receives pollen and is the tip of the **pistil**

stolon (botany): a creeping horizontal plant stem or runner that takes root at points along its length to form new plants

stochastic: random or involving chance

storm front: border between two different kinds of air, often with different temperatures; when a front is created, weather such as storms and **precipitation** happen

streamers (tail feathers): tail streamers on birds are the elongated narrow tips of the tail feathers; elongated, specialised tail feathers that are often used for display and communication, particularly during mating rituals. They can be quite long, sometimes exceeding the bird's body length, and are often vibrant in colour or pattern

streamline, streamlined: designed to have the least resistance

style (botany): of an angiosperm flower is an organ of variable length that connects the ovary to the stigma

sub- (prefix): underneath, below, less than

subcellular: of less than cellular scope or level of organisation

subpopulation: specific small section within a larger group

subspecies: additional species within a grouping

subtropical: bordering on the tropics; nearly tropical. pertaining to or occurring in a region between tropical and temperate; semitropical

superficial (anatomy): on the surface

superior (anatomy): above, higher than

supra- (prefix): above, beyond, or earlier in time

supralittoral: area above shoreline, likely damp

survive, survival: stay alive, staying alive

susceptibility: easily affected, harmed, or influenced by something

swamp: an area of low-lying, uncultivated ground where water collects; a bog or marsh

swamp drainage: emptying water of swamp

swim bladder: special organ inside a fish that helps it float and control its depth in the water

symbiotic, symbiosis: living together in a mutually beneficial way

tail streamers: tail streamers on birds are the elongated narrow tips of the tail feathers; elongated, specialised tail feathers that are often used for display and communication, particularly during mating rituals. They can be quite long, sometimes exceeding the bird's body length, and are often vibrant in colour or pattern

tannin: astringent vegetatation compound, reddish/brown

tassel, tasselled: fringed lobes of skin similar to tassels which are made of multiple cords, usually hanging down

tawny: dark or dull yellowish-brown

temperate: moderate in respect to temperature

termitarium/termitaria: termite nest/s

terrestrial: land based, growing or living on land

territory, territories, territorial: a tract of land, somewhere that the animal recognises as their territory which they will protect if they are territorial

thermal: using, resulting from, or producing heat

thermal current: energy that is transferred from one body to another because of a difference in temperature creating a current

thorax: the portion of the body between the head and the abdomen

threat: alarm, danger, hazard, menace, peril, risk, warning

threatened: in danger or likely to be harmed, or to become more significantly vulnerable or endangered

tissue: a group of similar cells and their intercellular material that work together to perform a function

tolerate: put up with

torpor: a state of physical or mental inactivity; lethargy.

totem, totem animal: a natural object or animal that is believed by a particular society to have spiritual significance and that is adopted by it as an emblem

toxic: poisonous

trample. trampled: when hard-hooved animals stamp or tread heavily on the ground

transmit, transmitted: to send or pass something from one place or person to another, like a message or a disease

transparent: see-through, light passes through completely like clear glass

transverse (anatomy): dividing body into upper and lower sections

trauma, traumatic: frightening, dangerous, or violent event that poses a threat to life or bodily integrity; wound or injury which can be both physical and/or psychological, or both

trawl: (myotis) fish by dragging feet through water

tree canopy: the upper parts of a tree, leaning into neighbouring trees

trenches (ocean zone): hadal zone

tropics, tropical: regions of Earth closest to the imaginary line called the equator, which runs around the middle of the planet. It's always warm in the tropics, with high humidity and plenty of rain

truffle: a fungus that grows underground and is edible

truncate: to cut off short; abbreviate, cut, foreshorten, shorten

trypanosome: genus of **kinetoplastids**: a monophyletic group of unicellular parasitic flagellate protozoa

tuber/s: underground part of a stem or rhizome, like potatoes or sweet potatoes.

tubercles, tuberculate: a little bump or knob, kind of like a tiny wart, that can grow on plants or on the body, especially on bones

tundra: major zone of treeless level or above the timberline on high mountains, rolling ground found in cold regions, mostly north of the Arctic Circle

turbulent/turbulence: characterised by conflict, disorder, or rapid, uneven changes. It can describe situations, periods of time, or even physical movements like air or water flows. Think of it as the opposite of smooth, stable, or calm.

tympanum: the eardrum or tympanic membrane

understorey: the layer of plants and especially the trees and shrubs between the forest canopy and the ground cover

unicellular: consisting of a single cell

unique, uniqueness: one of a kind, unlike anything else

urban: relating to a particular place, town,
village, in this instance development due to humans needing to have more homes

vagrant: wandering from place to place

vary, variable, variation: the quality or state of being different or diverse, not uniform

vegetarian: eating only plant matter

vegetation: plants

vehicular traffic: movement of cars, trucks, and other vehicles on roads

venom, venomous: poison, poisonous

venom glands: the specialised glands responsible
for producing the poisonous secretions of an animal, often associated with spines, teeth, stings, or other piercing devices

vent (anatomy): the external opening of the **cloaca**, a single posterior orifice that serves as the only opening for
the digestive, urinary, and reproductive tracts in many animals, including birds, reptiles, amphibians, and some fish and mammals

ventral (anatomy): towards the front

ventrolateral/ventrolaterally: the front and side of a body or body part

verify, verified, verification: prove the truth of, evidence

vertebra, vertebrae, vertebrate:
spinal column or backbone

vertical: upright, top to bottom

vicinity: the area around something

virus: an infectious microbe consisting of a segment of nucleic acid (either DNA or RNA) surrounded by a protein coat. A virus cannot replicate alone; instead, it must infect cells and use components of the host cell to make copies of itself

vivid: strikingly bright or intense

vulnerable: weak, unsafe, defenceless, sensitive

water hole: a natural puddle or small lake in a dry area where animals go to drink

water-logged: saturate with water

water table: the boundary between saturated and unsaturated ground

weed: plant growing where it's not wanted

weir/s: bank, barrier, ditch, dike

wet sclerophyll forest: characterised by very tall eucalypt trees usually in moist cloudy uplands with up to 2500mm of rainfall each year

wetlands: an area of land that's covered by water or has waterlogged soil, meaning the ground is full of water, like a sponge

whorl/s: something that swirls or winds around like a spiral

wingspan: the maximum extent across the wings of a bird or other flying animal, measured from tip to tip

xanthophore/s: a **chromatophore** containing a yellow pigment that is typically a **carotinoid** and occurring especially in fishes and crustaceans

REFERENCES AND CITATIONS:

1. IUCN (2024, July 19). *IUCN Red List*. International Union for the Conservation of Nature. Retrieved August 8, 2024, from https://www.iucnredlist.org/
2. Murray–Darling basin - Wikipedia. **https://en.wikipedia.org/wiki/Murray-Darling_basin**
3. MDBA: **https://images.app.goo.gl/5LzPYWJXLhgtZ8ct9**
4. **https://www.mdba.gov.au/water-management/infrastructure/lower-lakes-barrages**
5. **https://www.sciencedirect.com/science/article/abs/pii/S1462901121001453**
6. Kangaroo Island bee B&Bs aim to save native pollinators | Faculty of Sciences, Engineering and Technology | University of Adelaide. **https://set.adelaide.edu.au/news/list/2020/07/14/kangaroo-island-bee-bbs-aim-to-save-native-pollinators**
7. Australian masked owl - Wikipedia. **https://en.m.wikipedia.org/wiki/Australian_masked_owl**
8. River Red Gum forests and wetlands. **https://environmentvictoria.org.au/river-red-gum-forests-and-wetlands/**
9. Scarred tree - Wikipedia. **https://en.wikipedia.org/wiki/Scarred_tree**
10. Can Major Mitchell's cockatoo talk? - DIY Seattle. **https://diyseattle.com/can-major-mitchell-s-cockatoo-talk/**
11. Blue-winged Parrot - BirdLife Australia. **https://birdlife.org.au/bird-profiles/blue-winged-parrot/**
12. Stagonopleura guttata. **https://www.scientificlib.com/en/Biology/Animalia/Chordata/Aves/StagonopleuraGuttata01.html**
13. 20 Fun Koala Facts for Kids to Spark Classroom Conversation | Teach Starter. **https://www.teachstarter.com/us/blog/koala-facts-for-kids/**
14. Bettongs (Rat Kangaroos) | Bush Heritage Australia. **https://www.bushheritage.org.au/species/bettongs**
15. Boodie - Wikipedia. **https://en.wikipedia.org/wiki/Boodie**
16. Eastern quoll - Animal - Whois - xwhos.com. **https://www.xwhos.com/person/eastern_quoll-whois.html**
17. Eastern Quoll | Our Animals | Oakvale Wildlife. **https://oakvalewildlife.com.au/explore/our-animals/eastern-quoll**
18. Eastern barred bandicoot - Wikipedia. **https://en.wikipedia.org/wiki/Eastern_barred_bandicoot**
19. Black-tailed godwit (Limosa limosa). **https://www.thainationalparks.com/species/black-tailed-godwit**
20. Black-tailed godwit (Limosa limosa) - Wild travel in Danube Delta. **https://wildtravel.ro/bird/black-tailed-godwit-limosa-limosa/**
21. https://australian.museum/learn/animals/insects/macleays-swallowtail/
22. Far Eastern Curlew - BirdLife Australia. **https://birdlife.org.au/bird-profiles/far-eastern-curlew/**
23. https://biocache.ala.org.au/occurrences/47ef4b6a-3a1f-4ac8-801b-b1c3fd5df9b8

To claim your Conservation Club reward badge for THE MURRAY-DARLING BASIN AND THE COORONG environments:

- **Answer the following 20 questions.**
- **Send either a photograph or a scan of one colouring page** which you have coloured in reasonably correctly as far as the colours of your chosen species are concerned!

* **Postage will apply both in Australia and overseas. This will be estimated and sent to you for approval via email prior to shipping.**

Let's see what you have learned, whether you are 8, 80, or somewhere in between!

The answers can be found throughout the book.

1. How many species can you find in the book that are VULNERABLE on any of the IUCN, EPBC or NCA lists?
2. Which fish, featured in this book, has no scales and a species able to climb waterfalls?
3. Which insect digs its nests in soft wood and is currently part of a conservation project?
4. Which tiny bird is brave enough to make its nests in the base of the nests of large birds of prey?
5. Which seabird has long red tail streamers?
6. Which snake is also known as the "hoop snake" due to the defensive stance it takes on if threatened?
7. Which Australian marsupial is also known as a "rat kangaroo" and has a really unusual name?
8. Which butterfly feeds on and lays its eggs in citrus orchards, with the caterpillars looking a little bit like bird droppings?
9. Which spiky Australian critter has pink milk?
10. Which bird can be reunited with its mate within an average of three days when returning to its breeding grounds?
11. Which bird gets its name because it turns over stones when looking for food?
12. Which rainbow coloured bird beats bees against fenceposts to remove the stings?
13. Which bird is famous for its mating dance, and also has hybridized with the Sarus crane?
14. Which carnivorous bird is the second-largest nocturnal raptor in Australia?
15. Which avian species can travel almost twice as fast as a cheetah?
16. Which bird is more likely to be found by hearing its calls rather than sighting it?
17. Which marsupial can be distinguished from other species of this animal because it has only four toes on the hind feet?
18. Which marsupial makes V-shaped incisions in the bark of eucalypt trees to access sap?
19. Which marsupial with long ears is often modelled in chocolate and used in Easter baskets instead of Easter bunnies?
20. What type of plant, featured in this book, has a metallic, iridescent sheen to its flowers?

<u>FROG ID – Things to look for when trying to identify frogs....</u>

To be able to accurately identify frogs, you will need to develop some knowledge of their physical characteristics far beyond the information given here. These are just some very <u>basic clues</u> and are based on information from the book listed below, which, if you are a keen **herper** or likely to become one, I cannot recommend highly enough! This book narrows things down using a dichotomous key which you can go through step-by-step!

> **<u>Sanders MG (2021) *Photographic Field Guide to Australian Frogs.* CSIRO Publishing, Melbourne.</u>**

Head shapes:

> ***Dorsal outline:*** rounded, pointed, truncate
>
> ***Profile outline:*** rounded, sloping, bent, truncate, acute

Canthus shapes: rounded, angular, curved, and nearly straight

Pupil shapes: vertical, round, rhomboidal, horizontal

Skin textures: smooth, coarsely granular, finely granular, finaly granular with prominent tubercles and rounded ridges, irregularly textured, irregularly tuberculate, spinose and large rounded glands

Skin colour patterns: spots, blotches, ocelli, flecks, marbling, streaks and stripes, bands or bars

> **NB:** Sometimes temperatures or the substrate on which they are resting can influence the colours that they display – just to complicate things!

Frogs have four fingers and five toes!

> Some have long, skinny digits, others have shorter, stubby digits.
>
> Some have digits with terminal tips or swelling or 'pads', others do not have terminal digit swelling.
>
> Some have very distinctive webbing, others not so much - there is also variation in where the webbing occurs.

Location/distribution:

> A very good clue as to what many species would be and where they would occur is where they are found. This is beyond the scope of the information I can offer here.

I hope that these physical characteristics you can look for will help you narrow down the species you might have seen. Make sure to take notes of where you were (geographically and in terms of environments), what time of the year it was (seasons and temperatures could be important), and any other observations that you can note down in terms of descriptions. Take photos if you are able to!

ABOUT THE AUTHOR

Meet Elysia Rainwood, an artist and writer with a passion for conservation and environmental protection, particularly the preservation of the diverse animals inhabiting our wonderful Australian continent.

Growing up amongst the rustling leaves of eucalyptus trees, Elysia developed a passion for the unique creatures and the diversity of plant life found in Australia. Elysia always had an eye for detail, being very aware of such things as seasonal changes in the various environments observed. From the depths of the lush, green rainforest to the sun-kissed deserts, her childhood adventures fueled a desire to share the stories of the vulnerable and endangered.

Elysia writes non-fiction books and stories that resonate with the spirit of the tiniest of creatures and soar with the majesty of the wedge-tailed eagle. She likes to focus on the less well-known species, which may exist in your backyard, to ensure that they and their habitats are not overlooked.

Elysia Rainwood's mission is clear:
To enchant young minds with the wonders of Australia's wildlife, fostering a generation of conservationists. Through interesting facts, engaging and reasonably accurate illustrations, she hopes to ignite a spark in the hearts of 8 to 14-year-old readers, inspiring them to protect the precious creatures that share our planet.

Along the way, Elysia hopes to also engage with parents, care givers, educators who can pass on this information as well as those who just enjoy the relaxation that colouring careful illustrations can bring.

So, whether you are a youngster or someone who shares this passion for conservation, join Elysia on this informative journey into the magical realms of nature, where words become wings for the endangered and vulnerable to soar.

INDEX

Made in the USA
Monee, IL
08 November 2026

68085579R00103